Reliability of Nanoscale Circuits and Systems

Miloš Stanisavljević · Alexandre Schmid ·
Yusuf Leblebici

Reliability of Nanoscale Circuits and Systems

Methodologies and Circuit Architectures

Springer

Miloš Stanisavljević
Ecole Polytechnique Fédérale
de Lausanne
1015 Lausanne, Switzerland
milos.stanisavljevic@epfl.ch

Alexandre Schmid
Ecole Polytechnique Fédérale
de Lausanne
1015 Lausanne, Switzerland
alexandre.schmid@epfl.ch

Yusuf Leblebici
Ecole Polytechnique Fédérale de Lausanne
1015 Lausanne, Switzerland
yusuf.leblebici@epfl.ch

ISBN 978-1-4419-6216-4 e-ISBN 978-1-4419-6217-1
DOI 10.1007/978-1-4419-6217-1
Springer New York Dordrecht Heidelberg London

Library of Congress Control Number: 2010936070

Printed on acid-free paper

Springer is part of Springer Science+Business Media (www.springer.com)

There's Plenty of Room at the Bottom

Richard P. Feynman

Preface

The invention of integrated circuits and the continuing progress in their manufacturing processes are the fundamental engines for the implementation of semiconductor technologies that support today's information society. The vast majority of microelectronic applications presented nowadays exploit the well-established CMOS process and fabrication technology which exhibit high reliability rates. During the past few decades, this fact has enabled the design of highly complex systems, consisting of several millions of components, where each one of these components could be deemed as fundamentally reliable, without the need for extensive redundancy.

The steady downscaling of CMOS technology has led to the development of devices with nanometer dimensions. Future integrated circuits are expected to be made of emerging nanodevices and their associated interconnects. The expected higher probabilities of failures, as well as the higher sensitivities to noise and variations, could make future integrated circuits prohibitively unreliable. The systems to be fabricated will be made of unreliable components, and achieving 100% correctness of operation not only will be extremely costly, but may turn out to become impossible. The global picture depicts reliability emerging as one of the major threats to the design of future integrated computing systems. Building *reliable systems out of unreliable components* requires increased cooperative involvement of the logic designers and architects, where high-level techniques rely upon lower-level support based on novel modeling including component and system reliability as design parameters.

In the first part, this book presents a state of the art of the circuits and systems, architectures, and methodologies focusing on the enhancement of the reliability of digital integrated circuits. This research field spans over 60 years, with a remarkable revival in interest in recent years, which is evidenced by a growing amount of literature in the form of books, or scholarly articles, and comes as a reaction to an expected difficult transition from the CMOS technology that is widely perceived as very reliable into nanotechnology which is proven very unreliable in contrast. Circuit- and system-level solutions are proposed to overcome high defect density. Their performance is discussed in the context of a trade-off solution, where reliability is suggested as a design parameter to be considered in addition to the widely used triplet consisting of delay, area, and power.

Reliability, fault models, and fault tolerance are presented in Chapter 2, establishing the major concepts further discussed in the book. Chapter 3 depicts an overview of nanotechnologies that are considered in the fabrication of future integrated circuits. This work is focused at device level and addresses technologies that are still in relative infancy. Nanoelectronic devices prove to be very sensitive to their environment, during fabrication and operation, and eventually unreliable, thereby motivating the stringent need to provide solutions to fabricate reliable systems. Fault-tolerant circuits, architectures, and systems are explored in Chapter 4, presenting solutions provided in the early ages of CMOS, as well as recent techniques. Reliability evaluation, including historical developments, and also recent methodologies and their supporting software tools are presented in Chapter 5.

In the second part of the book, original circuit- and system-level solutions are presented and analyzed. In Chapter 6, an architecture suitable for circuit-level and gate-level redundant module implementation and exhibiting significant immunity to permanent and random failures as well as unwanted fluctuation of the fabrication parameters is presented, which is based on a four-layer feed-forward topology, using averaging and thresholding as the core voter mechanisms. The architecture with both fixed and adaptable threshold is compared to triple and R-fold modular redundancy techniques, and its superiority is demonstrated based on numerical simulations as well as analytical developments. Its applicability in single-electron-based nanoelectronics is analyzed and demonstrated.

A novel general method enabling the introduction of fault tolerance and evaluation of the circuit and architecture reliability is proposed in Chapter 7. The method is based on the modeling of probability density functions (PDFs) of unreliable components and their subsequent evaluation for a given reliability architecture. PDF modeling, presented for the first time in the context of realistic technology and arbitrary circuit size, is based on a novel reliability evaluation algorithm and offers scalability, speed, and accuracy. Fault modeling has also been developed to support PDF modeling.

In the third part of the book, a new methodology that introduces reliability in existing design flows is proposed. The methodology is presented in Chapter 8, which consists of partitioning the full system to design into reliability-optimal partitions and applying reliability evaluation and optimization at the local and system level. System-level reliability improvement of different fault-tolerant techniques is studied in depth. Optimal partition size analysis and redundancy optimization have been performed for the first time in the context of a large-scale system, showing that a target reliability can be achieved with low to moderate redundancy factors ($R < 50$), even for high defect densities (device failure rate up to 10^{-3}).

The optimal window of application of each fault-tolerant technique with respect to defect density is presented as a way to find the optimum design trade-off between the reliability and power area. R-fold modular redundancy with distributed voting and averaging voter is selected as the most promising candidate for the implementation in trillion-transistor logic systems.

The recent regain of interest in reliability that the community of micro and nanoelectronics researchers and developers shows is fully justified. The advent of novel methodologies enabling the development of reliable systems made of unreliable devices is a key issue to sustain the consumer and industry demands related to integrated systems with improved performance, lower cost, and lower power dissipation. This ultimate goal must be tackled at several levels of the VLSI abstraction, simultaneously, where the improvements at the lower levels provide benefits at the higher levels. Finally, also the upper levels including the compiler and software should be included in a common effort to reach this striving goal.

Lausanne
June 2010

Miloš Stanisavljević
Alexandre Schmid
Yusuf Leblebici

Acknowledgments

The authors would like to express their sincere thanks and appreciation to all the persons who helped during the course of writing this book.

The authors are grateful to the IEEE, Springer, ITRS for permission granted to reproduce some of the material from their earlier publications.

Heartfelt appreciation goes to the reviewers of the initial proposal of this book, Prof. Giovanni De Micheli, Prof. Kartik Mohanram, and Dr. Maria Gabrani, for investing their time to proofread and evaluate the research.

The editorial staff of Springer, especially Mr. Brett Kurzman, Editor Engineering Springer, has been highly supportive from the beginning of the project.

The authors gratefully acknowledge the support of the Swiss National Science Foundation and the Competence Centre for Materials Science and Technology.

Finally, we would like to acknowledge the support and invaluable encouragement of our families through the course of writing this book.

About the Authors

Miloš Stanisavljević received his M.Sc. degree in electrical engineering from the Faculty of Electrical Engineering, University of Belgrade, Belgrade, Serbia, in 2004, and his Ph.D. degree in electrical engineering from the Swiss Federal Institute of Technology (EPFL), Lausanne, Switzerland, in 2009. During 2004, he was an analog design and layout engineer for Elsys Design, Belgrade/Texas Instruments, Nice. In the end of 2004, he joined the Microelectronic Systems Laboratory, EPFL, as a research assistant. During 2006, he was with International Business Machines Corporation (IBM) Research, Zurich, for 6 months, where he was involved in a project related to reliability emulation in the state-of-the-art nanoscale CMOS technology. He is currently engaged in the field of reliability and fault-tolerant design of nanometer-scale systems. His current research interests include mixed-signal gate- and system-level design, reliability evaluation, and optimization. Dr. Stanisavljević received a Scholarship for Students with Extraordinary Results awarded by the Serbian Ministry of Education from 1996 to 2004.

Alexandre Schmid received his M.Sc. degree in microengineering and his Ph.D. degree in electrical engineering from the Swiss Federal Institute of Technology (EPFL) in 1994 and 2000, respectively. He has been with the EPFL since 1994, working at the Integrated Systems Laboratory as a research and teaching assistant and at the Electronics Laboratories as a post-doctoral fellow. He joined the Microelectronic Systems Laboratory in 2002 as a senior research associate, where he has been conducting research in the fields of non-conventional signal processing hardware, nanoelectronic reliability, bio-electronic and brain–machine interfaces. Dr. Schmid has published over 70 peer-reviewed journals and conference papers. He has served in the conference committee of The International Conference on Nano-Networks since 2006, as technical program chair in 2008, and general chair in 2009. Dr. Schmid is an associate editor of *IEICE ELEX*. Dr. Schmid is also teaching at the microengineering and electrical engineering departments/sections of EPFL.

Yusuf Leblebici received his B.Sc. and M.Sc. degrees in electrical engineering from Istanbul Technical University, in 1984 and in 1986, respectively, and his Ph.D. degree in electrical and computer engineering from the University of Illinois at Urbana-Champaign (UIUC) in 1990. Between 1991 and 2001, he worked as a faculty member at UIUC, at Istanbul Technical University, and at Worcester

Polytechnic Institute (WPI). In 2000–2001, he also served as the microelectronics program coordinator at Sabanci University.

Since 2002, Dr. Leblebici has been a chair professor at the Swiss Federal Institute of Technology in Lausanne (EPFL) and director of Microelectronic Systems Laboratory. His research interests include design of high-speed CMOS digital and mixed-signal integrated circuits, computer-aided design of VLSI systems, intelligent sensor interfaces, modeling and simulation of semiconductor devices, and VLSI reliability analysis.

He is the coauthor of four textbooks, namely *Hot-Carrier Reliability of MOS VLSI Circuits* (Kluwer Academic Publishers, 1993), *CMOS Digital Integrated Circuits: Analysis and Design* (McGraw Hill, 1st edition 1996, 2nd edition 1998, 3rd edition 2002), *CMOS Multichannel Single-Chip Receivers for Multi-Gigabit Optical Data Communications* (Springer, 2007), and *Fundamentals of High Frequency CMOS Analog Integrated Circuits* (Cambridge University Press, 2009), as well as more than 200 articles published in various journals and conferences.

He has served as an associate editor of *IEEE Transactions on Circuits and Systems* (II) and *IEEE Transactions on Very Large Scale Integrated (VLSI) Systems*. He has also served as the general co-chair of the 2006 European Solid-State Circuits Conference and the 2006 European Solid State Device Research Conference (ESSCIRC/ESSDERC). He is a fellow of *IEEE* and has been elected as distinguished lecturer of the *IEEE Circuits and Systems Society* for 2010–2011.

Contents

List of Figures

List of Tables

Acronyms

4LRA	four-layer reliable architecture
ADC	analog-to-digital converter
ADD	algebraic decision diagram
AES	advanced encryption standard
AFTB	atomic fault-tolerant block
AMC	airborne molecular contaminations
ANN	artificial neural network
ATC	averaging and thresholding circuit
ATPG	automated test pattern generation
BDD	binary decision diagram
BN	Bayesian network
C-SET	capacitive input SET
CA	cellular automata
CCC	custom configurable computer
CDF	cumulative distribution function
CED	concurrent error detection
CLB	configurable logic block
CMOL	CMOS-molecular electronics
CNN	cellular nonlinear network
CNT	carbon nanotube
CRC	cyclic redundancy check
CRMR	cascaded *R*-fold modular redundancy
CTL	capacitive threshold logic
CTMC	continuous time Markov chain
CTMR	cascaded triple modular redundancy
DAC	digital-to-analog converter
D2D	die-to-die
DCVS	differential cascode voltage switch
DES	discrete-event simulation
DIFTree	dynamic innovative fault tree

DTMC	discrete time Markov chain
ECC	error correcting code
EDA	electronic design automation
FA	full adder
FFANN	feed-forward artificial neural network
FO4	fanout of 4
FPGA	field-programmable gate array
GOS	gate oxide short
HARP	hybrid automated reliability predictor
HPTR	hardware partition in time redundancy
IEEE	Institute of Electrical and Electronics Engineers
IC	integrated circuit
IFA	inductive fault analysis
ITRS	International Technology Roadmap for Semiconductors
MCI-HARP	Monte Carlo integrated HARP
MC	Monte Carlo
MDP	Markov decision process
MRAM	magnetic random access memory
MTTF	mean time to failure
MVS	mid-value selection
NDR	negative differential resistance
NW	nanowire
PDE	partial differential equation
PDF	probability density function
PGM	probabilistic gate model
PLA	programmable logic array
PMC	probabilistic model checking
PRISM	probabilistic symbolic model checker
PTM	probabilistic transfer matrix
QCA	quantum cellular automata
QTR	quadruple time redundancy
RESO	recomputing with shifted operand
RESWO	recomputing with swapped operand
RETWV	recomputing with triplication with voting
RIR	R-fold interwoven redundancy
RMR	R-fold modular redundancy
RSFQ	rapid single flux quantum
RTD	resonant tunneling device
RTT	resonant tunneling transistor
RWPV	recomputing with partitioning and voting
SET	single-electron transistor
SEU	single-event upset
SE	soft errors
SHARPE	symbolic hierarch. automated reliability and perform. evaluator
SPRA	signal probability reliability analysis

TIR triple interwoven redundancy
TMR triple modular redundancy
TSTMR time shared triple modular redundancy
VHDL very-high-speed integrated circuits hardware description language
VLSI very-large-scale integration
WID within-die

Chapter 1
Introduction

1.1 From Microelectronics to Nanoelectronics

CMOS scaling has been an essence which has provided the semiconductor industry with historically unprecedented gains in productivity and performance, as depicted in Fig. 1.1, and which is proclaimed in the famous Moore's law [1]. Scaling has been the trend for decades and even though it has faced many barriers, clever engineering solutions and new device architectures have thus far broken through such barriers enabling scaling to continue at the same speed, and possibly at a slightly slower pace for the next 10 years. The nanoage has already begun, where typical feature dimensions are considered to be smaller than 100 nm. The operation frequency is expected to increase up to 12 GHz, and a single chip will contain over 12 billion transistors in 2020, as given by the International Technology Roadmap for Semiconductors initiative (ITRS) [2]. ITRS also predicts that the scaling of CMOS devices and process technology, as it is known today, will become much more difficult as the industry advances toward the 16 nm node and further.

With device geometries scaling below the 45-nm range, the available reliability margins are drastically reduced [4]. As a result, the reliability community is forced to thoroughly investigate accurate metrics which are able to determine these margins and how current reliability assessment methodologies must be adapted to gain a new reliability space for the most advanced technologies. Currently, from the chip designers' perspective, reliability increasingly manifests itself as time-dependent uncertainties of electrical parameters. In the sub-45 nm era, these device-level parametric uncertainties become too high to handle with prevailing worst-case design techniques, without incurring significant penalty in terms of area, delay, and energy consumption. Additionally, with continued scaling, the copper resistivity sharply increases due to interfacial and grain boundary scattering.

As the miniaturization trend approaches the physical limits of operation and manufacturing, the characterization of devices and circuit parameters becomes increasingly hard and even unpractical, with a lack of efficient solutions [5]. Due to the foreseeable limitations of silicon-based technology and the promising results of new devices of different nature working at nanometer level, there is a worldwide attention to the research and development of new electronic devices that can be the foundation of the future integrated circuit fabrication technology.

M. Stanisavljević et al., *Reliability of Nanoscale Circuits and Systems*,
DOI 10.1007/978-1-4419-6217-1_1,

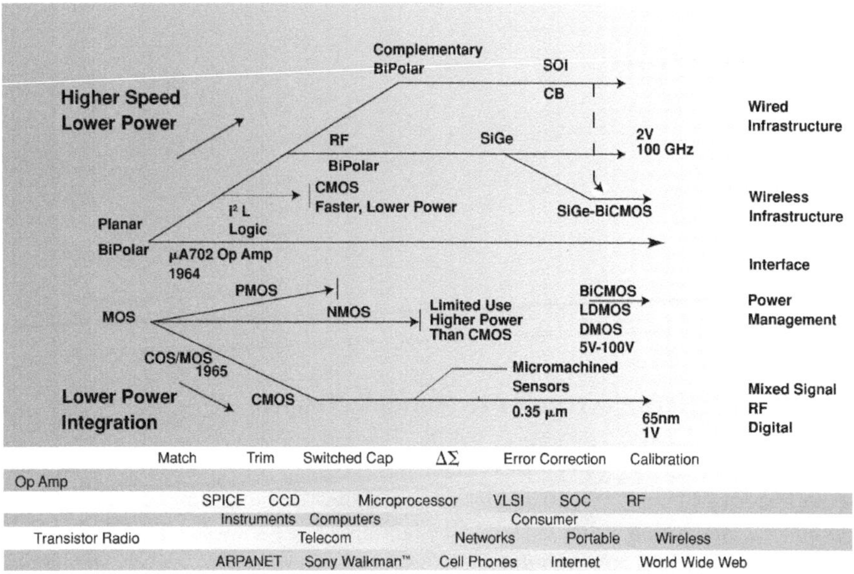

Fig. 1.1 Brief history of the semiconductor industry (adapted from [3], © [2007] IEEE)

Future systems based on non-CMOS nanoelectronic devices are expected to suffer from low reliability, due to permanent and transient errors. The permanent error rate will increase due to constraints imposed by fabrication technologies. The transient errors rate will increase due to nondeterministic parasitic effects such as background charge, which may disrupt correct operation of single devices in both time and space, in a random way. Higher operating frequencies pose strict limits to timing and therefore also introduce the probability of timing errors.

The increased integration of devices on a single die raises the probability of erroneous components within a die. The individual device failure rates also increase. The manufacturing failure rate per device for present-day CMOS technology is approximately in the range of 10^{-7}–10^{-6} [6]. The expected probability of failure during manufacturing of nanoscale or molecular scale devices will be several orders of magnitude higher [7–10].

The probability of failure can be directly associated with the yield. In the semiconductor industry, yield is represented by the functionality and reliability of integrated circuits produced on the wafer surfaces. During the manufacturing of integrated circuits yield loss is caused by defects, faults, process variations, and design. With constant scaling the impact of each individual factor on yield is increasing which is illustrated in Fig. 1.2.

During processes as implantation, etching, deposition, planarization, cleaning, lithography, etc., failures responsible for yield loss are observed. Causes and mechanisms responsible for yield loss are numerous: (a) airborne molecular contamination (AMC) or particles of organic or inorganic matter caused by the environment or by the tools; (b) process-induced defects as scratches, cracks, and particles, overlay

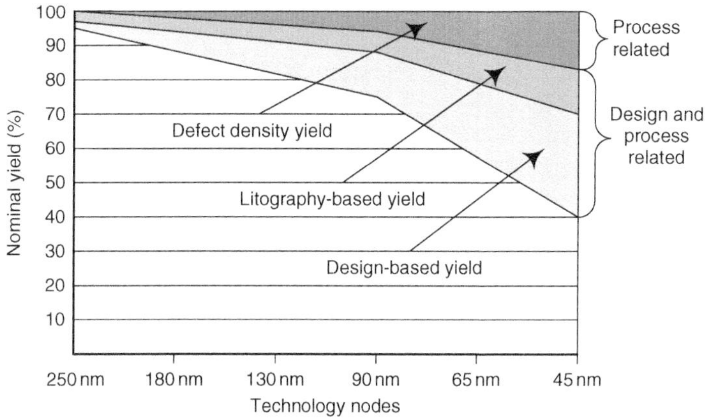

Fig. 1.2 Impact of different factors on yield, over technology scaling (adapted from [11] according to data from [12])

faults, and stress; (c) process variations resulting, e.g., in differing doping profiles or layer thicknesses; (d) deviation from design, due to pattern transfer from the mask to the wafer, resulting in deviations and variations of layout and critical dimensions; and (e) diffusion of atoms through layers and in the semiconductor bulk material. The most common causes of defects are illustrated in Fig. 1.3. The determination of defects and yield and an appropriate yield to defect correlation are essential for yield enhancement [12].

The study of fault tolerance as we know it today emerged from the science of information theory (or informatics). The well-known approaches for developing fault-tolerant architectures in the presence of uncertainties (both permanent and transient faults) consist of incorporating redundancy [13]. Even though these fault-tolerant methods perform efficiently in the context of low failure density that have been encountered so far, the massive nature of defects expected to plague early generations of nanometric devices demands fundamentally original approaches to be applied. For example, triple modular redundancy (TMR) used with majority voting has established as a major reliability enhancement technique for systems based on CMOS devices. TMR is applied to large systems or components, typically in the scale of computer in space and aircraft systems. However, TMR is only efficient for defect densities in the range of 10^{-8}–10^{-7} [6], which disqualifies TMR for systems consisting of unreliable nanodevices. Advanced fault-tolerant strategies must be developed in order to enable accommodating more than 10^{12} molecular-sized devices on a 1 cm^2 chip [7].

The unprecedented amount of computational power that these new technologies are expected to permit will only be exploitable if new design methodologies are available. The main reasons for this are related to the huge complexity of such systems and the high number of defective components that unavoidably will appear with the introduction of emerging and future technologies. Consequently, the

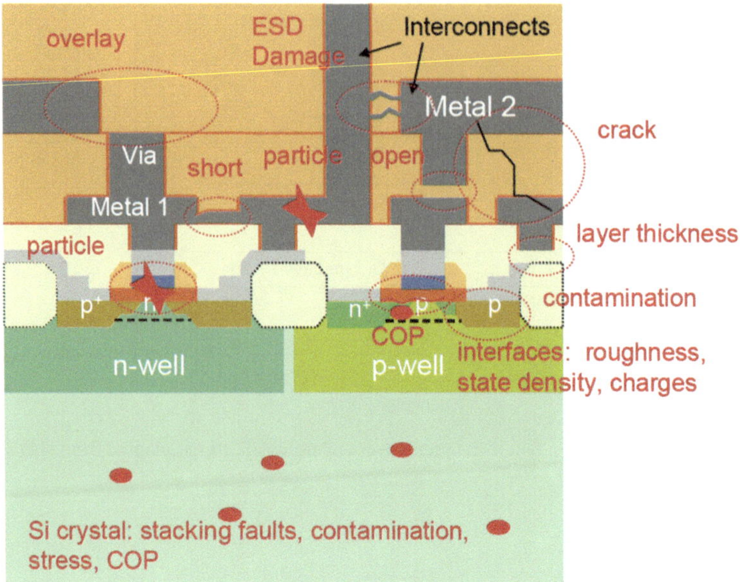

Fig. 1.3 Various types of defects in integrated circuits (Semiconductor Industry Association. The International Technology Roadmap for Semiconductors, 2009 Edition. SEMATECH: Austin TX, 2009 [2])

expected panorama of future electronic system design methodologies corresponds to a massive use of components, in orders of magnitude higher than today's, with component reliabilities in orders of magnitude lower than today's. This represents a new, challenging, and essential problem. Nowadays, the design strategy is based on the hierarchical characterization of several levels of abstraction, from device to architectural high level, involving intrinsic verification methods and tools at each level. This allows the treatment of large circuits at different abstraction and complexity levels. In this scenario, the designer assumes that final systems are composed of perfect or acceptably correct components. Designers are only aware of a potential defect through the use of design for testability rules, tools, and standards, which aims at making the last test manufacturing control stage that separates good and bad circuits simple and efficient. While the vast majority of recent nanoelectronics-related research efforts concentrate on the development of new nanomaterials and devices, very little has been accomplished into the direction of design methodologies for circuits and systems using such emerging technologies. The main reasons behind this trend are as follows: (i) the perception that the novel device technologies are still immature to justify any exploration of design methodologies; (ii) the assumption that once the new devices are available, one can utilize well-known design paradigms, methodologies, and tools in a straightforward manner to develop circuits and systems; and (iii) the reluctance of industry to figure out and solve problems that are expected to affect the next generation of circuits [14].

1.2 Issues Related to Reliable Design

Several issues need to be addressed in the development of original and efficient fault-tolerant methods, in regard to the specific new constraint related to the necessity of operating under massive defect density, i.e., proper system operation must be guaranteed if several functionally redundant units are defective, and also considering the need to limit the redundancy factors to very low numbers. We believe that the following issues and visions need to be addressed in order to provide future microelectronic systems with functional robustness:

- Systems to be fabricated will be made of unreliable components. How reliable systems are fabricated from unreliable components represents a major issue to be solved. Moreover, the probabilistic nature of the component reliability has to be accepted as a new parameter; the design process will have a stochastic component, as an immediate consequence.
- Maintaining 100% correct operation in the presence of high defect density is not only very expensive in terms of area and power, but might be plainly impossible. Hence, relaxing the requirement of 100% correctness for devices and interconnects may reduce the costs of manufacturing, verification, and test [15]. Where today's approach consists of extensive testing prior to shipping fully reliable components, we think that future microelectronic systems will be fabricated with a variable degree of robustness, reflecting how much effort has been placed into the introduction of hardware for fault tolerance.
- Reliability is becoming an important parameter that needs to be included in new design methodologies, and must become the fourth optimization pillar of the nanoelectronics, along with the well-known triplet power, area, and speed [16].
- Fault tolerance of future nanoelectronic circuits must be handled jointly at several levels of abstraction, e.g., at circuit level, architectural level, as well as at system level and at the algorithmic level; the massive nature of fault density demands cooperative efforts from these various levels to absorb or recover from faults [14].
- The support for a priori estimation of the required redundancy and the optimal size of partitions and reliable blocks to be inserted into the hardware, with respect to the desired reliability must be provided, taking into account realistic failure models for several types of disruptions, to correct transistor operation.

To offer solutions to some of the aforementioned issues, we can perceive some important tasks and prerequisites that need to be fulfilled:

- proposing and verifying a new fault-tolerant architecture that enables improvement in reliability, with respect to the existing ones;
- building a very realistic fault model, which is relevant for further evaluation of the reliability;
- developing tools and methods for accurate reliability evaluation of existing and new fault-tolerant architectures;

- performing an analysis of existing and new fault-tolerant architectures at system level in order to identify an optimal window of operation with respect to the defect density and extracting the most promising architecture under study;
- performing a system-level analysis to acquire reliability-optimal redundancy factors and partition sizes;
- performing reliability-optimal partitioning at system and local levels;
- integrating the proposed tasks into a new fault-tolerant design methodology that should be an "upgrade" of the existing digital design methodology and merge into it in a seamless and transparent way.

1.3 Outline of the Book

This book aims at providing a wide overview of reliability within the context of submicron CMOS and nanoelectronic developments and present solutions to related reliability issues. The organization of the book is as follows. Chapter 2 introduces terms of reliability and fault tolerance as well as faults and fault modeling. A realistic transistor fault model for existing CMOS technology intended to be used in reliability simulations is introduced. In Chapter 3, a detailed overview of future nanotechnologies and nanodevices such as single-electron transistors (SET), nanowires, carbon nanotubes (CNT), crossbars, quantum cellular automata (QCA) is presented. Defects and fault rates in nanotechnologies are also examined. All popular fault-tolerant techniques including static (modular redundancy, NAND/MAJ multiplexing), dynamic (error-correction codes), and hybrid techniques are presented in Chapter 4. Different concepts and tools for reliability evaluation as well as a Monte Carlo tool developed by the authors are presented in Chapter 5, including detailed description of algorithms and realizations. In Chapter 6, a novel fault-tolerant technique based on averaging and adaptable thresholding and including redundancy is proposed, and various implementations are discussed. They include the evaluation of the reliability of the proposed fault-tolerant technique for standard CMOS gates and small circuit, for differential logic and future nanodevices, such as the single-electron transistor. Chapter 7 presents a novel general method enabling the introduction of fault tolerance and evaluation of circuit and architecture reliability. The method is based on the modeling of probability density functions (PDFs) of unreliable components and their subsequent evaluation for a given reliability architecture. In Chapter 8, a new methodology that introduces reliability in existing design flows is presented. The methodology consists of a priori system-level reliability evaluation and optimization, reliability-optimal partitioning, and local-level reliability evaluation. System-level reliability improvement of different fault-tolerant techniques is analyzed in depth, and the optimal window of application of each fault-tolerant technique with respect to defect density is derived, along with the optimal redundancy factor and partition size. Finally, a concluding discussion is presented in Chapter 9.

Chapter 2
Reliability, Faults, and Fault Tolerance

A clear understanding of several concepts and terminology related to reliability is needed to proceed with the understanding of the methodologies which are applied to guarantee optimal operability of VLSI systems, fault tolerance, and circuit architectures implementing them. Basic terms such as reliability, fault tolerance, faults, and fault modeling are introduced and explained in detail. The chapter is organized as follows. In Section 2.1, the general concepts of reliability, fault tolerance, and yield are explained. Faults and fault models are presented in Section 2.2. A realistic transistor fault model adapted to current CMOS technology is presented in Section 2.3.

2.1 Reliability and Fault Tolerance

Reliability is defined according to *IEEE* as the ability of a system or component to perform its required functions under stated conditions and for a specified period of time. The process yield of a manufacturing process is defined as the fraction, or percentage, of acceptable parts among all parts that are fabricated [17]. A system failure occurs or is present when the *service* provided by the system differs from the specified service or the service that should have been offered. In other words, the system *fails* to perform what it is expected to.

In classical theory [18, 19], the reliability $R(t)$ is defined as the probability of a system to operate correctly during the time interval $[0,\ t]$, given that it has been operative at time 0. Let $F(t) = P\{T \leq t\}$ be the probability that a failure occurs at a time T, smaller than or equal to t, then

$$F(t) = \int_{-\infty}^{t} f(t)dt, \tag{2.1}$$

where $f(t)$ represents the probability density function (PDF) of the random variable, time to failure. $R(t)$ represents the probability that a system has not failed by time t, which is expressed as $R(t) = P\{T > t\}$, and consequently

DOI 10.1007/978-1-4419-6217-1_2,

$$R(t) = 1 - F(t). \tag{2.2}$$

The failure rate λ represents the probability that a failure occurs within a time interval $[t_1, t_2]$, given that it has not occurred prior to t_1. In electronic systems, λ can legitimately be considered constant, and in this case,

$$R(t) = e^{-\lambda t}. \tag{2.3}$$

Finally, the mean time to failure (MTTF) is expressed as the expected value of time to failure and is derived as

$$\mathrm{MTTF} = \int_0^\infty R(t)dt, \tag{2.4}$$

and upon constant failure rate,

$$\lambda = \frac{1}{\mathrm{MTTF}}. \tag{2.5}$$

The so-called bathtub curve which is shown in Fig. 2.1 is widely accepted to represent a realistic model of the failure rate of electronic equipment and systems over time [20]. The bathtub curve consists of three characteristic zones. Failure rates follow a decreasing pattern during the early times of operation, where infant mortality deteriorates the system, typically due to oxide defects, particulate masking defects, or contamination-related defects. Failure rate remains constant over the major part of the system operation life. Failures are random, mostly manifesting themselves as soft errors. Wearout occurs in the final stage of the system lifetime, where failure rate increases, typically due to electromigration-related defects, oxide wearout, or hot carrier injection.

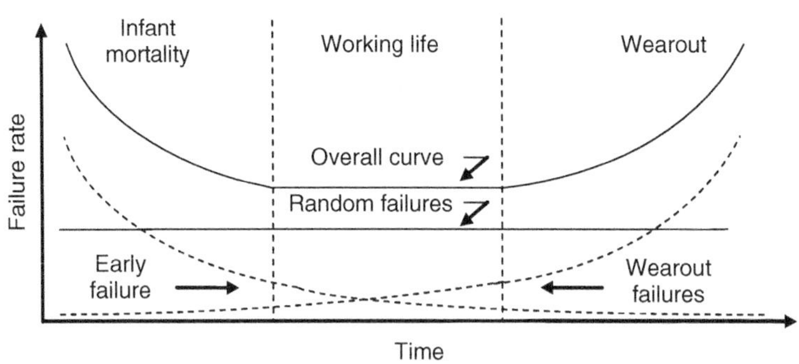

Fig. 2.1 Bathtub curve; the time axis is not to scale ([21], with kind permission of Springer Science and Business Media). The hard curve shows cumulative contributions of its three components that are presented as the dotted curves named Early failure and Wearout failures and the solid line named Random failures

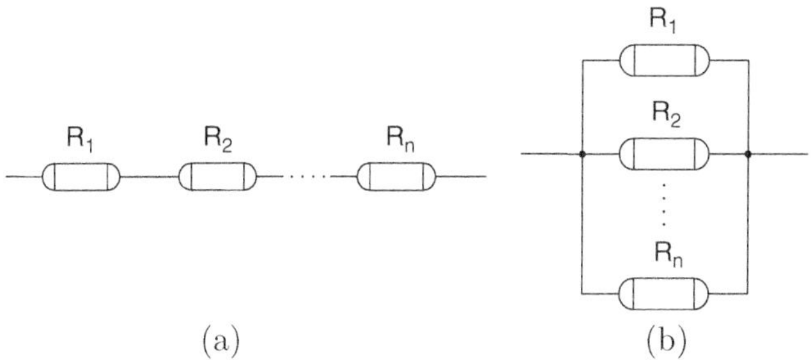

Fig. 2.2 Electrical component configurations: **(a)** serial and **(b)** parallel

Some major architectural configurations of electronic systems are very common, and the analysis of their reliability behavior forms the foundation of the analysis of any complex system. In the serial configuration, depicted in Fig. 2.2a, several blocks, n, with failure rates $R_1(t), \ldots, R_n(t)$ considered independent of each other are cascaded. The correct operation of the system depends on the reliability of each block and is mathematically expressed as

$$R_{\text{system}} = R_1(t) \cdot R_2(t) \cdot \cdots \cdot R_n(t) = \prod_{i=1}^{n} R_i(t). \tag{2.6}$$

In the parallel configuration, depicted in Fig. 2.2b, malfunction of all composing blocks is necessary to cause the system to fail. Naming the probability of failure or unreliability of the components $Q_i = 1 - R_i$ and omitting the expression of time (t) for clarity, the probability of failure of the system is expressed as

$$Q_{\text{system}} = \prod_{i=1}^{n} Q_i. \tag{2.7}$$

The reliability of the system composed of parallel implementation is expressed as

$$R_{\text{system}} = 1 - Q_{\text{system}} = 1 - \prod_{i=1}^{n}(1 - R_i) \tag{2.8}$$

and can be higher than the reliability of individual components if redundancy is applied. Realistic designs are typically composed of hybrid arrangement of parallel and serial configurations, where the system reliability can be obtained by iterative decomposition of the network into its series and parallel components and step-by-step solving.

Finally, a system in a k-out-of-n configuration consists of n components. Only k components need to function properly to enable the full system to operate.

A system which has the ability to deliver the expected service operation despite the occurrence of faults or the presence of defects is named fault tolerant. Fault tolerance of microelectronic systems is presented in detail in Chapter 4.

2.2 Faults and Fault Models

The following three terms are crucial and related to system failure and thus need to be clearly defined, which are named *defect*, *error*, and *fault* [17].

A defect in an electronic system is the unintended difference between the implemented hardware and its intended design. Some typical defects of VLSI chips include [22]

- process defects, taking the form of missing contact windows, parasitic transistors, oxide breakdown, etc.;
- material defects, due to bulk defects (cracks, crystal imperfections), surface impurities, etc.; and
- age defects, taking the form of dielectric breakdown, electromigration, etc.

Defects can be also classified by the statistical effect they produce:

- Systematic, defects that have the same impact across large dimensions, such as die or wafer, and that can be modeled in a systematic way. These defects are usually the result of process–design interaction.
- Random (stochastic), all types of defects that cannot be controlled or modeled in a predictable and systematic way. They include random particles in the resist or in the materials, inserted or removed, or defects in the crystal structure itself that alter the intended behavior of the material and results in excessive leakage or in a shift in the device threshold (V_{th}), eventually causing the failure of the device. The failure modes resulting from these defects are

  1. Opens
  2. Shorts
  3. Leakage
  4. V_{th} shift
  5. Variability in mobility (μ)

Random defects do not necessarily result in a complete failure of the device, but in a significant deterioration of its performance [23].

Some classical microphotographs of defects are presented in Fig. 2.3. The patterns are easily recognizable and are presented as illustrative cases. Visual inspection cannot be applied in the detection of defects in modern digital systems, consisting of hundreds of billions of transistors and their interconnections routed over nine metal layers. Test techniques are applied [17], which form a discipline of its own. The application of test techniques and enabling the testability of complex digital systems (design for testability) imposes the designers additional constraints in terms of classical circuit specifications (area, delay). In addition, any fault-tolerant technique

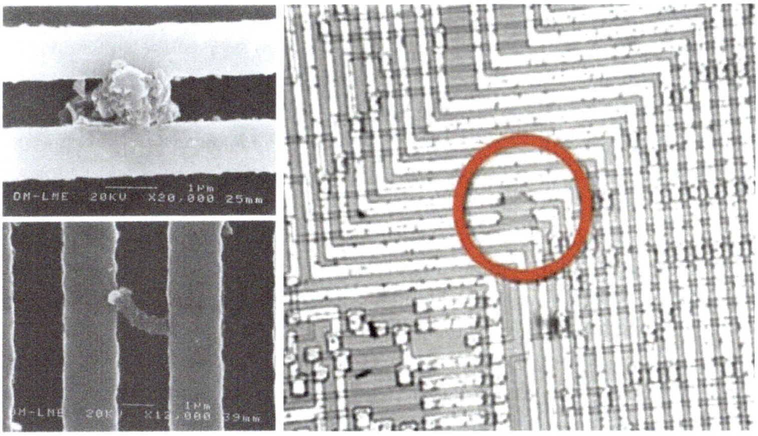

(a) (b)

Fig. 2.3 Defect images. **(a)** Bridging defects with low-resistance electrical behavior on the *top* and high-resistance electrical behavior on the *bottom* microphotograph ([24], with kind permission of Springer Science and Business Media) and **(b)** open defect inside the *circle* ([25], with kind permission of Springer Science and Business Media)

which is implemented at the hardware level must be proven compliant with the test methodologies, which may be a difficult task.

The existence or emergence of defects reduces yield.

A wrong output signal produced by a defective system is called an error. An error is an effect whose cause is some defect. Errors can be classified into three main groups, namely permanent, intermittent, and transient errors, according to their stability and concurrence [26].

- Permanent errors are caused by irreversible physical changes in a chip. The most common sources for this kind of errors are the manufacturing processes. Permanent errors can also occur during the usage of the circuit, especially when the circuit is old and starts to wear out. Common to all permanent errors is that once they have occurred, none will vanish and consequently the test to detect them can be repeated, conducting to identical results. Permanent errors are also known under the denomination of hard errors.
- Intermittent errors are occasional error bursts that usually repeat themselves every now and then and are not continuous as permanent errors. These errors are caused by unstable or aging hardware and are activated by an environmental change such as a temperature or voltage change. Intermittent errors often precede the occurrence of a permanent error; for instance an increased resistance in a wire may be observed before it cracks, creating an open circuit. Intermittent errors are very hard to detect because they may only occur under certain environmental constraints or in the presence of some specific input vector combination.
- Transient errors are temporal single malfunctions caused by some temporary environmental conditions which can be an external phenomenon such as radiation or noise originating from other parts of the chip. Transient errors do not leave any

permanent marks on the chip and therefore they are also called soft errors (SE). A common manifestation of transient error is a change of the binary value of a single bit (e.g., a bit flip in memory cell). Another term, single-event upset (SEU), is used for soft error, which describes the fact that malfunctions (upsets) are commonly caused by single events such as an absorbed radiation. The occurrence of transient errors is commonly random and therefore hard to detect.

Error sources can be classified according to the phenomenon causing the error. Such origins are for instance related to the manufacturing process, physical changes during operation, internal noise caused by other parts of the circuit, and external noise originating from the chip environment.

A fault is a representation of a defect at the abstracted functional level. A fault is present in the system when physical difference is observed between the "good" or "correct" system and the actual system. Discussions presented in this book mostly relate to permanent faults caused by physical defects.

The most common faults in a chip are spots and bridging faults caused by silicon impurities, lithography, and process variations [27]. These faults cause permanent errors in a circuit. The probability of these defects is likely to increase with technology scaling, as larger numbers of transistors are integrated in a single chip and the size of chips increases, while devices and wires sizes decrease. This results in a decreasing yield, and consequently higher price, of functioning chips. The move toward nanoscale circuits also raises a list of new problems originating from the manufacturing process. As the fabrication dimensions shrink, the proportional extent of deviations becomes larger and their effects more severe. Lithography deviation is the main cause of gate length deviations. Moreover, fluctuations of the doping profile in turn cause deviations of the transistor threshold voltage. These effects, together with the increase of resistive vias and contacts, eventually result in large operation speed deviation. Simultaneously, the operation frequency of integrated circuits is expected to increase. The worst-case scenario consisting of a series configuration of "slow" devices may lead to timing violations and therefore to malfunction of the circuit. This is considered an intermittent error because the circuit may correctly operate most of the time; this would not be the case for permanent errors.

The diminishing value of reliability of very deep submicron technologies is an established fact. Moreover, it is widely accepted that nanoelectronic-based systems will rely on a significantly lower reliability rate than what was known so far. More details of challenges and faults in nanodevices are given in the following chapter.

If error detection and recovery do not take place in a timely manner, a *failure* can occur that is manifested by the inability of the system to provide a specified service. *Fault tolerance* is the capability of a system to recover from a fault or error without exhibiting failure. A fault in a system does not necessarily result in an error; a fault may be latent in that it exists but does not result in an error; the fault must be sensitized by a particular system state and input conditions to produce an error. The techniques related to fault-tolerant systems include fault avoidance, fault masking, detection of erroneous or compromised system operation, containment of error propagation, and recovery to normal system operations [28].

Actual defects in a circuit cannot be directly considered in the design and validation of the circuit and therefore special fault models are needed. Fault models are simplifications of the phenomena caused by defects on the circuit and were first introduced by Eldred in the late 1950s [29]. Fault models have been developed at each level of abstraction, i.e., behavioral, functional, structural, switch, and geometric levels.

In this book, we limit our discussions to switch-level and geometric fault models. The higher level abstraction models do not offer a satisfying level of accuracy, which is required to study and apply fault-tolerant techniques further assessed. This comment also covers *stuck-at* (permanent connection of the gate input or the output to supply lines) and *von Neumann* fault models (which consists of transient bit-flip faults at the gates and interconnects [13]) which belong to structural fault models. Even though the stuck-at fault model is the most popular and widely used model in industry, which has the ability to detect a majority of physical defects, it is not adequate for accurate reliability evaluation in modern technologies [30, 31]. These referenced papers show that approximating the gate probabilities of failure by (bounding) constants introduces sizable errors, leading to overdesign. Moreover, stuck-at fault models will not be suitable for future nanodevices as demonstrated on the example of single-electron transistor (SET) circuits by Beiu et al. [31].

Switch-level fault models are defined at the transistor level. The most prominent fault models in this category are the *stuck-off/stuck-open* and *stuck-on/stuck-short* fault models. If a transistor is permanently in non-conducting state due to a fault, it is considered to be stuck-off or stuck-open. Similarly, if a transistor is permanently in conducting state, it is considered to be stuck-on or stuck-short. These fault models are specially suited for the CMOS technology.

Geometric fault models assume that the layout of the chip is known. For example, knowledge of line widths, inter-line and inter-component distances, and device geometries is used to develop these fault models. At this level, problems related to the manufacturing process can be detected. The layout information, for example, can be used to identify lines or components that are most likely to be shorted due to process defects. The *bridging* fault model leads to accurate detection of realistic defects. With shrinking geometries of VLSI chips, this model becomes increasingly important.

A new model for CMOS technologies that combines the benefits of switch-level and geometric fault models has been developed and is presented in Section 2.3. The model exhibits much better accuracy than typical switch-level models, while exhibiting a complexity comparable to switch-level models. Moreover, a simple fault model for SET has been developed and is used in simulations and results presented in Section 6.4.

2.3 Transistor Fault Model

A major step in any design automation process consists of simulation. In order to perform a simulation for reliability, an accurate and realistic fault model is necessary. Considering permanent errors as the main and most intricate source of unreliability, physical defects and fault modes are modeled with a netlist fault description. There are various ways of modeling physical defects, at various levels of abstraction, as presented in Section 2.2. Geometrical models that are close to the physical layout are complex and impractical in large-scale simulations. However, they are the most accurate. Statistical models related to physical defects distribution are not hard to embed into circuit-based analysis. The stuck-at approach which is traditionally used in fault coverage analysis is not sufficient to handle the analysis of various faults in nanometer-scale devices. The following two basic approaches are a starting point for our model, namely inductive fault analysis (IFA) [32] and transistor-level fault modeling [33], both of which have complex implementations.

The transistor-level fault modeling is applied at an abstraction level above the physical layout and can be classified as a switch-level fault model. It usually incorporates only stuck-on and stuck-off models of transistors for representing faults. These models represent a very reduced set of possible physical defects and therefore they are not sufficient. On the other hand, the IFA approach, which is a geometric fault model, has some drawbacks, mainly related to high computational complexity of the used tools, complete dependency on geometrical characteristics, and difficulty of properly handling analog layouts. Our model provides better accuracy that is comparable to IFA models and is operated with a time complexity comparable to switch-level models.

A hierarchical transistor fault model is developed in order to overcome shortfalls of transistor-level fault modeling using some results from the IFA approach and also to cover a range of impacts as wide as possible that device faults have on the circuit behavior. The fault model consists of two layers (Fig. 2.4). The upper layer (LY2) models various physical defects such as missing spot, unwanted spot, Gate oxide short (GOS) with channel, floating gate coupled to a conductor, and bridging

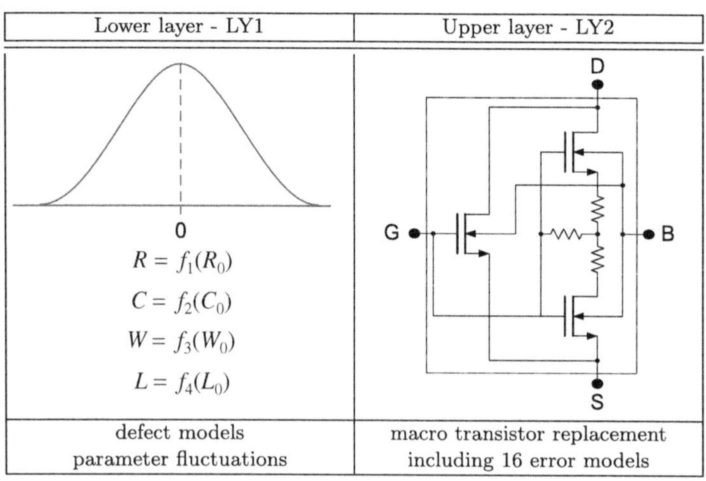

Fig. 2.4 Two-layer fault model

faults [34, 35]. Some of the physical defects are depicted in Fig. 2.3. The models have been developed from structural and lithography defects, and each defect model is described in terms of electrical parameters of its components. Thus, for simulation purposes, physical defects are translated into equivalent electrical linear devices such as resistors, capacitors and nonlinear devices such as diodes and scaled transistors. A total of 16 possible defects are considered for each transistor, which are listed in Table 2.1. The number of implemented defective transistor equivalent circuits is nine, while seven of them are available in two implementations, i.e., for high and low values of defect model parameters. All defective transistor equivalent circuits (for open drain, open source, floating gate, drain–source short, drain–gate short, gate–source short, drain–bulk short, source–bulk short, and gate oxide short) are depicted in Fig. 2.5a–i. Opens and shorts are modeled as a resistance which is placed in parallel with a capacitance on the spot of a defect [33, 36]. The floating gate (Fig. 2.5c) is modeled as a capacitive divider between the gate terminal and source [35, 37]. Gate oxide short (GOS; Fig. 2.5i) is modeled by dividing the gate area into three equivalent transistors: two are in a series configuration and are placed in parallel with the third one, with a common node at the location of the physical gate oxide short spot [35, 38].

Table 2.1 List of transistor failures modeled in the upper layer (LY2)

Acronym	Failure type
DHO	*D*rain *H*ard *O*pen, resulting in stuck-off fault
DSO	*D*rain *S*oft *O*pen, resulting in partial stuck-off fault
SHO	*S*ource *H*ard *O*pen, resulting in stuck-off fault
SSO	*S*ource *S*oft *O*pen, resulting in partial stuck-off fault
FLG	*FL*oating *G*ate resulting in disconnected input
DSHS	*D*rain *S*ource *H*ard *S*hort, resulting in stuck-on fault
DSSS	*D*rain *S*ource *S*oft *S*hort, resulting in partial stuck-on fault
DGHS	*D*rain *G*ate *H*ard *S*hort, resulting in input–output bridging fault
DGSS	*D*rain *G*ate *S*oft *S*hort, resulting in partial input–output bridging fault
GSHS	*G*ate *S*ource *H*ard *S*hort, resulting in input stuck-at fault
GSSS	*G*ate *S*ource *S*oft *S*hort, resulting in partial input stuck-at fault
DBHS	*D*rain *B*ulk *H*ard *S*hort, resulting in excessive current flowing through the substrate
DBSS	*D*rain *B*ulk *S*oft *S*hort, resulting in partial excessive current flowing through the substrate
SBHS	*S*ource *B*ulk *H*ard *S*hort, resulting in current flowing through the substrate only for non-common sources
SBSS	*S*ource *B*ulk *S*oft *S*hort, resulting in small current flowing through the substrate only for non-common sources
GOS	*G*ate *O*xide *S*hort, resulting in an excessive current flowing through the gate oxide insulator

The lower abstraction model layer (LY1) consists of defective transistor circuit model parameters (e.g., resistances R, capacitances C, and geometric parameters gate length L, gate width W for gate-oxide short model) whose variation can have a significant influence on the defect model. Here, each parameter is modeled with the Normal distribution – $\mathcal{N}(\mu, \sigma)$, with a nominal mean value (μ) and a given

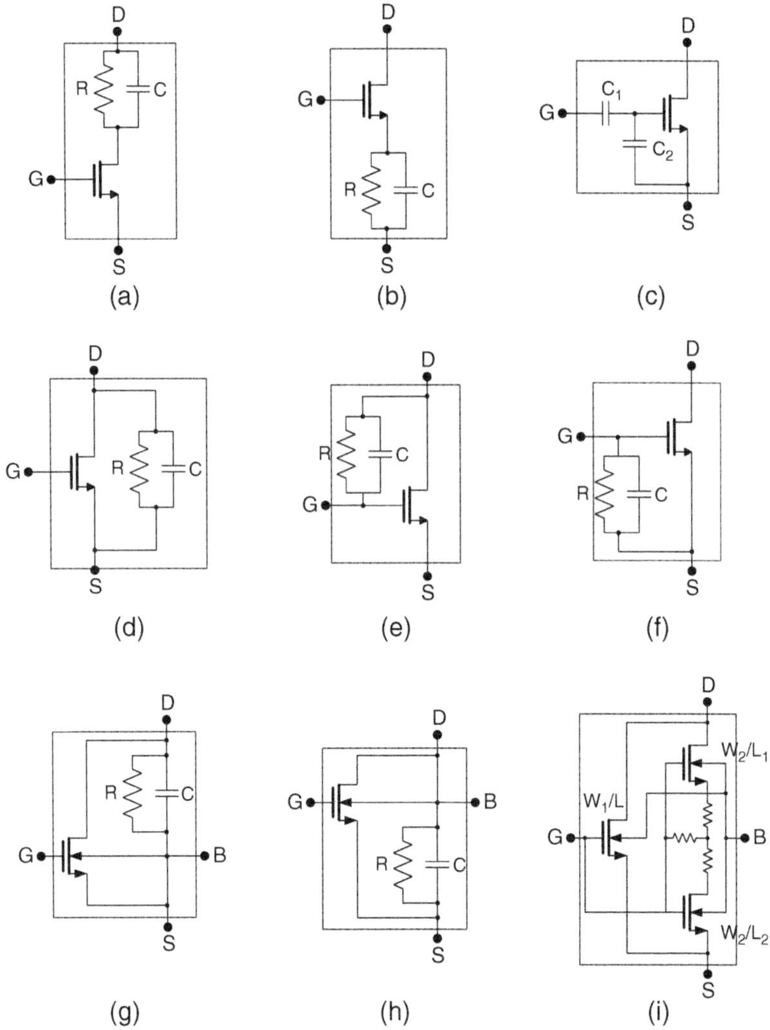

Fig. 2.5 Transistor equivalent defect models: (**a**) open drain, (**b**) open source, (**c**) floating gate, (**d**) drain–source short, (**e**) drain–gate short, (**f**) gate–source short, (**g**) drain–bulk short, (**h**) source–bulk short and (**i**) gate oxide short

standard deviation (σ). Nominal parameter values of R have been chosen according to [35, 37] to 1 and 5 kΩ resistance for hard and soft short defects, respectively, and to 100 and 0.5 MΩ for hard and soft opens, respectively.

An extraction of actual or realistic values of these parameters requires an access to the fabrication process parameters and test parameters that are usually kept confidential within the process manufacturer. However, some of the parameters may be extracted by means of building and measuring different testing structures on test chips. Some results have been presented in the comprehensive literature related to bridging faults [36, 37], resistive opens and shorts [39], and transistor gate geometrical parameters [40]. One possible test structure for extracting drain/source open resistance is illustrated in Fig. 2.6 and consists of an array of multiple transistors connected in series, and uniformly distributed over the chip, with the possibility of measuring the current flowing through each line. Here, IDDQ testing (which relies on measuring the supply current (I_{DD}) in the quiescent state) with the respective data from the process manufacturer regarding the probability of drain/source open could provide a means of extracting the nominal value of the resistance parameter.

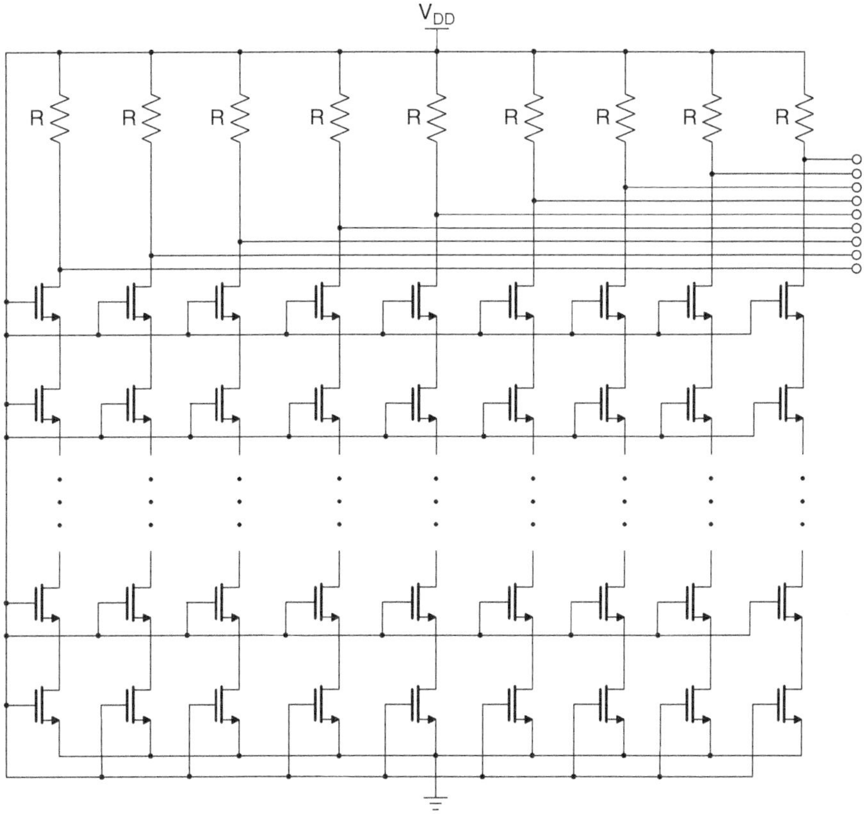

Fig. 2.6 Test structure for measuring drain/source open resistance parameter

The layer that represents the mapping of interconnection defects into their electrical models (open spots and bridging faults) [36] is not included in the defect models and simulations. Modeling of interconnection defects at system level is highly dependent on geometrical characteristics of the layout, where maintaining the correspondence between the physical and electrical parameters remains a problem that needs to be solved. In the transistor-level simulations, this layer can be excluded, considering that more than 80% [41] of signal errors in modern circuits are due to global signals stuck-at supply or ground.

The transistor-level model presented in this section will be widely used in reliability simulations throughout the course of this book.

Chapter 3
Nanotechnology and Nanodevices

The end of the ITRS roadmap for classical CMOS devices and circuits envisions the emergence of future nanotechnologies and nanodevices and also evidences many new related challenges. This chapter covers some of these issues using a tutorial presentation style.

Logic design at present is solely applied to microelectronics. The process of transfer of circuits and systems to nanoelectronics and relevant hybrid technologies (e.g., molecular electronics) has already started. Very fundamental and technological differences between nanoelectronic devices and microelectronic devices exist, those latter possibly in the nanometer size domain. Even though CMOS devices reach below 50 nm dimensions, these devices rely on enhanced but standard CMOS fabrication processes, and hence do not formally classify as nanoelectronic devices. Novel physics, integrated with design methods and nanotechnology, leads to far-reaching revolutionary progress. The main classification and the roadmap of some existing nanoelectronic devices is presented in Fig. 3.1, where the technology status is presented on the vertical axis.

Clearly, a significant amount of research is needed to step from the current state of the technology where the operation of single devices or basic Boolean gates is demonstrated toward the ultra-large integration comprising billions of individual devices and their interconnects, as well as a systematic hierarchical organization into architectural levels.

Figure 3.2 presents a partial taxonomy of nanoscale CMOS devices. A variety of prospective devices is under research. Over the years, new candidates to the replacement of the MOSFET emerged while some were abandoned. Generally, a number of devices appear promising and research is progressing toward the implementation of relatively basic Boolean structures such as gates to a full adder or simple analog structures such as ring oscillators. Memory systems are also considered, since their systematic arrangement naturally lends themselves to the implementation of arrays of nanodevices. Fabrication and modeling are progressing in parallel. In some cases, system-level simulators have been developed, enabling the early assessment of the prospects of a device and in the expectation that wide-range nanosystems are technologically viable in the near future.

M. Stanisavljević et al., *Reliability of Nanoscale Circuits and Systems*,
DOI 10.1007/978-1-4419-6217-1_3, © Springer Science+Business Media, LLC 2011

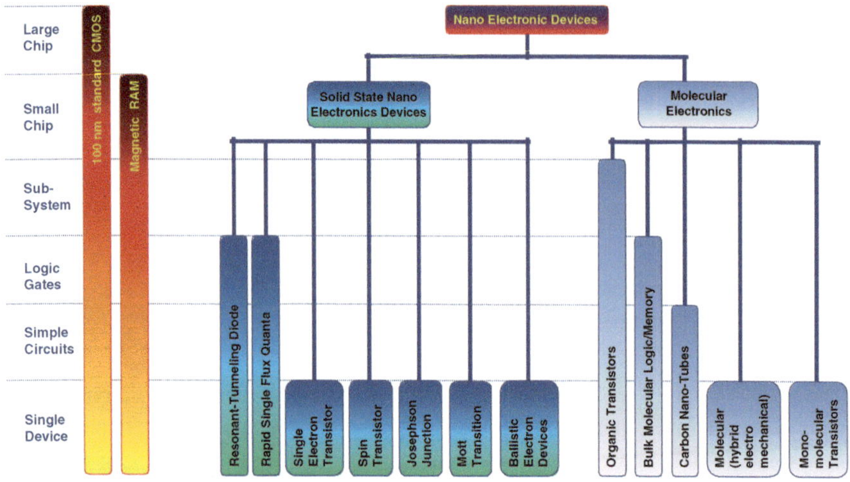

Fig. 3.1 The roadmap for nanotechnology presents many nanodevices currently being investigated as an alternative to standard CMOS (adapted from [42], © [2007] IEEE)

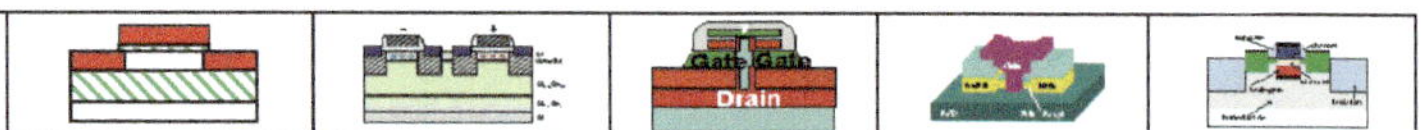

DEVICE	ULTRA-THIN BODY SOI	BAND-ENGINEERED TRANSISTOR	VERTICAL TRANSISTOR	FINFET	DOUBLE-GATE TRANSISTOR
CONCEPT	Fully depleted SOI	SiGe or Strained Si channel; bulk Si or SOI	Double-gate or surround-gate structure (No specific temporal sequence for these three structures is intended)		
APPLICATION/DRIVER	Higher performance, Higher transistor density, Lower power dissipation				
ADVANTAGES	-Improved subthreshold slope -V_t controllability	-Higher drive current -Compatible with bulk and SOI CMOS	-Higher drive current -Lithography independent L_g	-Higher drive current -Improved subthreshold slope -Improved short channel effect -Stacked NAND	-Higher drive current -Improved subthreshold slope -Improved short channel effect -Stacked NAND
SCALING ISSUES	-Si film thickness -Gate stack -Worse short channel effect than bulk CMOS	-High mobility film thickness, in case of SOI -Gate stack -Integration	-Si film thickness -Gate stack -Integrability -Process complexity -Accurate TCAD including QM effect	-Si film thickness -Gate stack -Process complexity -Accurate TCAD including QM effect	-Gate alignment -Si film thickness -Gate stack -Integrability -Process complexity -Accurate TCAD including QM effect
DESIGN CHALLENGES	-Device characterization -Compact model and parameter extraction	-Device characterization	-Device characterization -PD versus FD -Compact model and parameter extraction -Applicability to mixed signal applications		
MATURITY	Development				
TIMING	Near Future →				

Fig. 3.2 Nanoscale CMOS devices

In the following sections, a brief overview of some typical nanoelectronic and hybrid devices is presented, such as

- single-electron transistors (SETs),
- resonant tunneling devices (RTDs),
- quantum cellular automata (QCA),
- one-dimensional (1D) devices,
- CMOS-molecular electronics (CMOL), and
- other nanoelectronic devices such as rapid single flux quantum (RSFQ), superconducting circuits of Josephson junctions, and spin transistors.

In addition, design hurdles, demands of future nanoscale circuits, and considerations regarding faults in nanodevices are presented at the end of Section 3.8.

3.1 Single-Electron Transistors (SETs)

Single-electron tunneling devices (SETs) are three-terminal devices where electron movement through the device is controlled with a precision of an integer number of electrons. An electron can tunnel from and to an island or quantum dot, through a tunneling barrier, which is controlled by a separate gate based on Coulomb blockade. This electron island can accommodate only an integer number of electrons. This number may be as high as a few thousands. A single-electron transistor is composed of a quantum dot connected to an electron source and to a separate electron drain through tunnel junctions, with the electron injection controlled by a gate electrode. Single-electron transistors can be implemented in logic circuits by operating on one or more electrons as a bit of information [7].

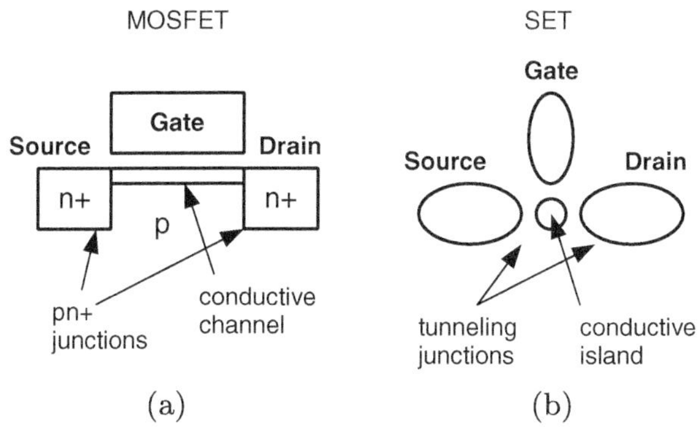

Fig. 3.3 Simplified structure of a MOSFET (**a**), compared with that of a SET (**b**)

The simplified structure of a SET is compared with that of a MOSFET in Fig. 3.3. Indeed, the device is reminiscent of a typical MOSFET, but with a small conducting island embedded between two tunnel barriers, instead of the usual inversion channel. The current–voltage characteristics of the SET are shown in Fig. 3.4, as

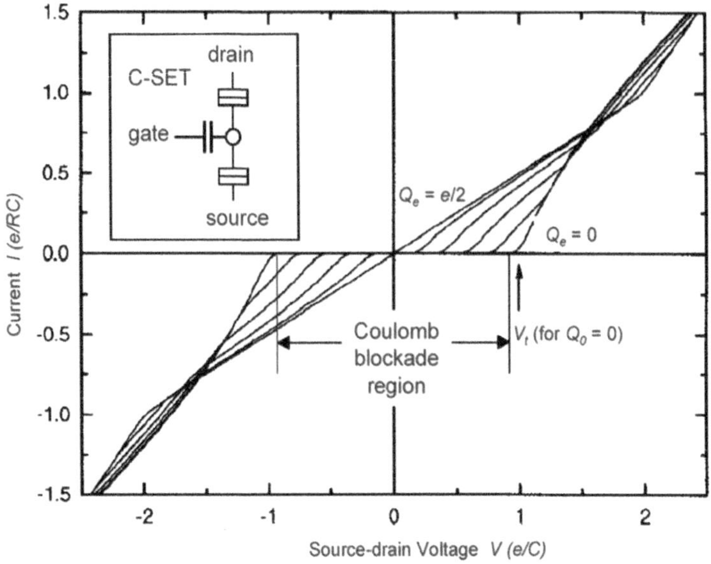

Fig. 3.4 Typical current–voltage characteristics of a C-SET displaying the Coulomb blockade region for low source–drain voltage values (adapted from [7], © [1999] IEEE)

a function of different gate voltage levels. At small drain-to-source voltages, there is no current since the tunneling rate between the electrodes and the island is very low. This suppression of DC current at low voltage levels is known as the Coulomb blockade. At a certain threshold voltage, the Coulomb blockade is overcome, and for higher drain-to-source voltages, the current approaches one of its linear asymptotes. A very significant property of the single-electron transistor is related to the fact that the threshold voltage and the drain-to-source current in its vicinity are periodic functions of the gate voltage. The physical reason for this periodicity lies in the fact that the conditions that govern the tunneling of charge between the electrodes and the isolated island can be established for consecutive, discrete states that correspond to the existence of integer multiples of an electron charge on the island. Still, it is evident that the device can be operated as a switch controlled by the gate electrode, capable of performing logic functions.

The dimensions of the conductive island and the tunneling junctions need to be in the order of a few nanometers to a few tens of nanometers. While larger device dimensions allow observable device operation at very low temperatures, the dimensions may need to be reduced to sub-nanometer levels in order to achieve Coulomb blockade near room temperature [7]. It is estimated that the maximum operation temperature for 2 nm SETs is 20 K, with an integration density of approximately 10^{11}cm^{-2} and an operating frequency in the order of 1 GHz [44]. Various logic applications of SETs, including inverters [45–48], OR, NAND, NOR, and a 2-bit adder [48–51], have been demonstrated. However, due to the high impedance required for Coulomb blockade, a SET gate would not be able to drive more than one cascaded gate. This has two implications. First, SET logic would have to be based on local architectures, such as cellular arrays and cellular nonlinear networks (CNNs) [2]. Second, although SETs may not be suitable for implementations in logic circuits, they could be used for memories. SET-based memory structures have been proposed and experimentally demonstrated [52–54].

Background charge fluctuations remain a major issue for the successful operation of SET-based circuits [7]. Due to electrostatic interactions, correct device function can be prevented by impurities and trapped electrons in the substrate. In order to tackle this problem, besides the endeavor to develop novel computing schemes, such as the multi-value SET logic, fault-tolerant architectures, implemented at higher levels of circuits and systems, may be a direction of investigation [55].

3.2 Resonant Tunneling Devices (RTDs)

Resonant tunneling devices form a well-known group of devices in the development of nanoelectronics. Today, resonant tunneling transistors (RTTs) are among the most established nanoscale devices since they operate at room temperature. Moreover, from the viewpoint of circuit applications their fabrication and interfacing with FETs and BJTs has reached an advanced level that allows the investigation of small-scale circuit development.

Resonant tunneling devices are usually two-terminal devices made of vertical semiconductor heterostructures with two insulating layers separating the conducting regions. A negative differential resistance (NDR) is produced by the double barrier structure, which has a resonance peak enabling the resonant tunneling of electrons through the barriers. Due to the fast tunneling process, RTDs inherently have a very high switching speed of up to 700 GHz, which makes them potentially attractive for high-speed switching applications, such as very high -frequency oscillators, amplifiers, and ADCs [2]. Three-terminal devices have been demonstrated by integrating RTDs with conventional FETs (RTD-FETs) [56]. Various designs, including digital logic, threshold logic, and memory, have been proposed based on the heterostructures of RTD-FETs [57, 58]. However, the combination of RTDs and transistors introduces delays to the intrinsically fast switching speed of RTDs. The operating speed of hybrid devices can be one order of magnitude slower than the switching speed of RTDs. Furthermore, the complexity of the integrated structure imposes a limit on the scaling properties of the devices, compared with CMOS. Resonant tunneling transistors have been obtained by adding a control terminal to the RTD [59] and RTT-based logic circuits have been demonstrated [60]. The compatibility with the fabrication process of silicon structures has also been demonstrated.

A major problem of RTDs lies in the extreme sensitivity of the device characteristics to the layer thickness, as the tunneling current exponentially depends on the thickness of the tunnel barrier. Difficulties in manufacturing enabling the large-scale production of RTD circuits with uniform thickness of tunnel barriers remain. Another problem is the low on/off current ratio of 10, which is far from the 10^5 ratio obtained with CMOS circuits [61]. These and other challenges in fabrication may limit the usefulness of RTDs to niche applications related to high-speed switching, digital signal processing, ADC, DAC, etc.

3.3 Quantum Cellular Automata (QCA)

Cellular automata (CA) are computing architectures inspired by complex natural and physical systems [62]. CA systems are usually based on regular arrays of simple cells. Each cell in an array interacts with its nearest neighbors and evolves from an initial state into a final state following a predefined rule. The evolution of a cell is determined by the cell's initial state and the interactions with its neighbors. A computation can be mapped to such a dynamic process in a CA system.

The quantum cellular automata (QCA) is a device concept and an architecture concept, which represent a new approach for information processing. QCA devices can be divided into three different categories, namely molecular QCAs, magnetic QCAs, and electrostatic QCAs. The concept of quantum cellular automata (QCA) was first proposed as a cell structure of quantum dots coupled via quantum mechanical tunneling [63]. The QCA basic block (electrostatic QCA) is a cell containing quantum dots that can be aligned in different ways representing binary information. Data are electrostatically propagated along the cells that can be arranged in two-dimensional arrays to perform logic operations or functions defined in the cellular automata theory (and cellular neural networks) [64].

In a typical four-dot cell, the quantum dots are placed in the corners of a square cell. Due to electrostatic repulsion, free charges will occupy the dots in diagonally opposite corners of the cell and form two bistable states representing binary bits. Logic states are thus encoded in the spatial distribution of electric charges in a cell and a computation can be performed by the mutual interactions of cells in an array. Basic logic circuits [65], a latch [66], and shift registers [67] have been experimentally demonstrated for electronic QCA implementations. Figure 3.5 shows the representation of QCA cells with four and six quantum dots.

The potential advantages of QCA are high switching speed, low power consumption, and good scaling capability. It is estimated that the inter-dot distance in a solid-state QCA cell would be approximately 20 nm and the inter-cell distance would be 60 nm [68]. In a recently proposed scheme for a molecular QCA cell [69], the inter-dot distance is expected to be about 2 nm and the inter-cell distance about 6 nm. An optimistic evaluation shows that the intrinsic switching speed of an individual QCA cell can be in the terahertz range [68].

However, it was shown by a comparative study of QCA and CMOS circuit performance that a practical circuit of solid-state QCA only has a maximum operating speed of a few megahertz [70]. This frequency may reach a few gigahertz for circuits based on molecular QCA. It was also shown that the maximum operating

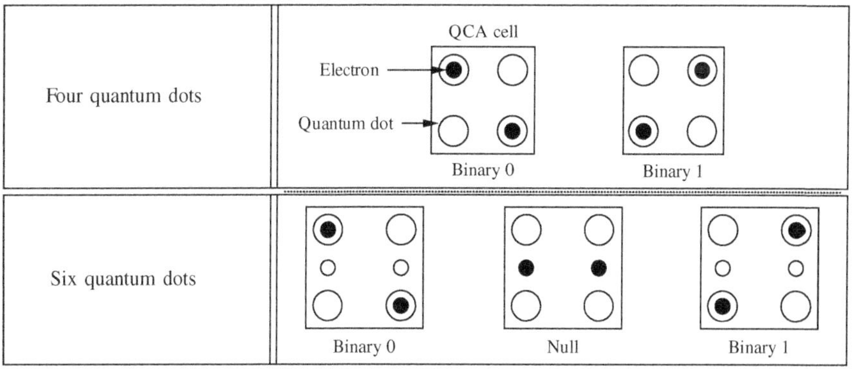

Fig. 3.5 QCA cells with four and six quantum dots

temperature for a standard solid-state QCA cell is approximately 7 K, indicating that room-temperature operation is not possible for solid-state QCA systems [68]. Other serious limitations of QCA devices are synchronization complexity (adiabatic clocking field) and the problem of background charge fluctuation, because QCA are single-electron devices. In the 2005 ITRS roadmap, the electrostatic QCA devices were not included in the tables because they are slow, they need low temperatures, and their applications are different from the ones of interest in the semiconductor industry. Molecular QCA systems may be the only possibility to obtain room-temperature operation and they appear in the molecular devices category.

Besides the widely studied electronic QCA, the concept of magnetic QCA based on small ferromagnetic structures has been proposed for room-temperature operation [71]. For magnetic QCA, logical states are represented by the directions of the cell magnetization and cells are coupled through magnetostatic interactions. The minimum size of magnetic QCA cells is estimated to be approximately 100 nm, and the maximum switching speed is approximately 200 MHz. Logic devices including a shift register have been demonstrated for the use of nanoscale ferromagnetic devices [72]. The properties researched in ferromagnetic logic are the nonvolatility and reconfigurability [2].

3.4 One-Dimensional (1D) Devices

While being referred to as one-dimensional (1D) devices, according to [2], carbon nanotubes (CNTs) and semiconductor nanowires (NWs) are often also considered as molecular devices. The potential advantages of 1D structures include enhanced mobility and phase-coherent transport of the electron wavefunctions. These properties may lead to faster transistors and novel wave interference devices. Carbon nanotubes and semiconductor nanowires are important subsets of 1D structures.

Carbon nanotubes are cylindrical structures made of rolled up graphite sheets (graphene) [64, 73]. Depending on the orientation (chirality) of the graphene forming the tube, the structure may have semiconductor or metallic properties. The tubes can be doped to construct p–n junctions. CNTs have some interesting properties, such as a high electrical and thermal conductivity, high tolerance to chemical corrosion and electromigration, and can sustain much higher currents than metals [73]. CNT dimensions may vary from 1 to 20 nm in diameter and from 100 nm to micrometers in length. CNTs have been studied in FET structures (CNT-FET) where the silicon channel of the transistor is replaced by a CNT. Figure 3.6a shows a CNT-FET.

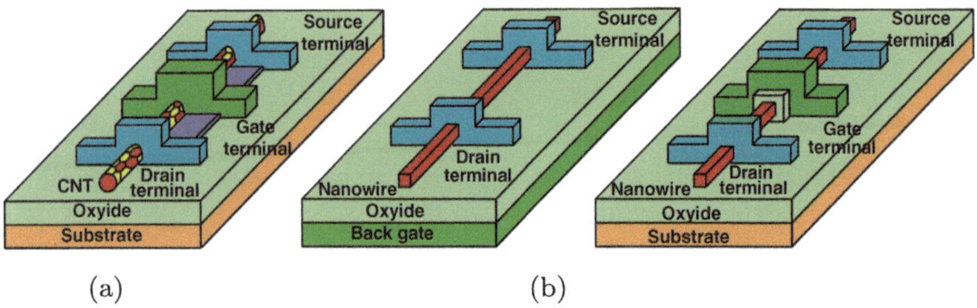

Fig. 3.6 1D structures: **(a)** CNT-FET; **(b)** two alternate nanowire transistor devices

Transistors have been obtained from CNTs [74, 75], and logic circuits, such as NOT, NOR, a flip-flop, and ring oscillators, have been demonstrated [76, 77]. However, it is still not possible to precisely control whether CNTs are semiconducting or metallic, which makes the fabrication of CNTs a process subject to random components.

The main challenges associated with CNTs are the non-deterministic chirality, placement and size of the fabricated tubes, and the high value of the contact resistance [64] that limits the maximum current flowing through the device. Recent research results show improvements in the precision of placement and in the control of chirality of CNTs [2].

Nanowires can be used in individual transistor structures or in array/crossbar structures [64]. When used as the channel element connecting source and drain, the characteristics of nanowires are better than those of bulk silicon in terms of switching speed. When used in array structures, the resistance of the crossing points of nanowires can be configured, and architectures such as programmable logic arrays (PLAs) can be implemented. Such array structures are conceptually simple, can achieve high density, and can be fabricated through a directed assembly process [78, 79]. A nanowire, usually with a diameter of 10–20 nm, can be doped as a p- or n-type device. NW FETs have been obtained by making structures of crossed p- and n-type nanowires separated by a thin dielectric [80]. Figure 3.6b shows transistor structures based on nanowires.

Various logic gates, also exhibiting gain, have been demonstrated [81]. More complicated circuits such as address decoders have recently been reported [82].

These results present a step toward the realization of integrated nanosystems based on semiconductor NWs.

Even though 1D structures (CNTs and NWs) are the most promising alternatives to hybrid integration with CMOS technology, some problems remain unsolved. These problems include the low drive capability of individual devices, their contact resistance limited by quantum effects, their interconnect problems (control of manufacturing and placement), and yield of fabrication.

3.5 CMOS-Molecular Electronics (CMOL)

Molecular electronics refers to devices where the switching or storage capacity is based on the operation of single molecules as basic building blocks [83]. Organic and inorganic molecular circuits are being researched to produce two- and three-terminal devices and the necessary interconnections. Considering the dimensions involved, molecular devices promise very high densities, increased switching speeds, and reduced energy consumption [56].

Logic circuits based on two-terminal devices and programmable molecular switches [43] have been experimentally realized. A three-terminal FET structure based on a C-60 molecule has been demonstrated, but exhibiting a very high contact resistance [84]. Along with conventional logic architectures, molecular electronics are suitable for integrating crossbar structures. The most elaborate molecular circuit available to date is a 64-bit random access memory, which has been experimentally realized on a two-dimensional (2D) crossbar circuit [85].

Large-scale molecular circuits can, in principle, be fabricated through self-assembly and a low-cost stochastic chemical or biological process, solving the increasing problem of nanoscale lithography. However, many technological challenges remain in building large-scale molecular circuits [86]. For example, no or very low gains are possible in molecular circuits, and most molecular devices have low on/off current ratios, which make molecular devices sensible to perturbations and noise. The problems of low yield in fabrication and reliability in operation due to the stochastically self-assembly process indicate that molecular computer systems would require defect- and fault-tolerant architectures for reliable operations.

The main problem today resides in synthesizing molecules that would combine suitable device characteristics with the ability to self-assemble, with high yield, and enabling a few nanometer gaps between pre-fabricated nanowires. The general idea of CMOS/nanowire/MOLecular hybrid circuits (depicted in Fig. 3.7) consists of combining the advantages of the currently dominating CMOS technology (including its flexibility and high fabrication yield) with those of molecular devices with nanometer-scale footprint. Two-terminal molecular devices would be self-assembled on a pre-fabricated nanowire crossbar fabric, enabling very high functional density at acceptable fabrication costs. However, CMOL technology imposes substantial requirements on circuit architectures, most importantly a high defect tolerance.

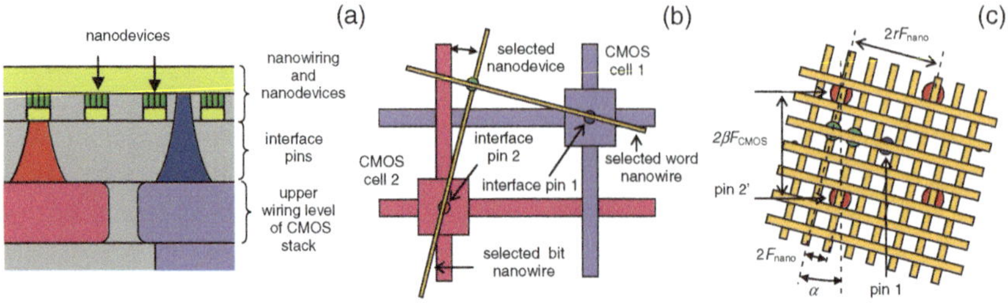

Fig. 3.7 Low-level structure of generic CMOL circuit: (**a**) a schematic side view; (**b**) a schematic top view showing the idea of addressing a particular nanodevice via a pair of CMOS cells and interface pins; and (**c**) a zoom-in top view on the circuit near several adjacent interface pins. On panel (**b**), only the activated CMOS lines and nanowires are shown, while panel (**c**) shows only two devices. (In reality, similar nanodevices are formed at all nanowire crosspoints.) Also disguised on panel (**c**) are CMOS cells and wiring (adapted from [87], © [2005] IEEE)

Although some progress has been made in the research of molecular devices, many challenges related to the molecular operation and molecular manufacturing remain unsolved at this point. Whether and when this technology will form a viable replacement or complement to CMOS technology is not clear.

3.6 Other Nanoelectronic Devices

Rapid single flux quantum (RSFQ) devices are based on the effect of flux quantization in superconducting Josephson junctions [88]. Josephson junctions serve as switching elements and binary bits are represented by the presence or absence of flux quanta in the superconducting circuits. A voltage pulse is generated when a magnetic flux quantum is transferred from one circuit to another by switching the Josephson junctions. Complex circuit functions are realized by the propagation and interaction of the voltage pulses in RSFQ circuits. Current RSFQ devices are mainly built using low-temperature superconductors (~5 K), while high-temperature superconductor (~50 K) technology may eventually be possible for implementations of RSFQ circuits.

The main advantage of RSFQ circuits lies in the very high operating speed, reaching up to approximately 770 GHz, which has been achieved in flip-flop circuits [89]. More complex circuits, such as random access memories, adders, and multipliers, have been demonstrated [90]. As the superconducting quantum effect occurs at a microscopic scale, the typical dimension of RSFQ devices is of the order of a few microns. It has been shown that RSFQ circuits can be scaled down to 0.3 μm and can operate at frequency of 250 GHz [91]. However, further scaling of RSFQ devices into nanoscale will be a challenge, due to many limiting factors associated with this technology, such as the magnetic penetration depth.

The main drawback of the RSFQ technology is the necessity of cryogenic cooling [92]. A broad scale of applications will strongly depend on the availability of low-cost, highly reliable, and compact cooling systems. Before significant technical progress is made in the development of cryogenic coolers, the RSFQ technology is likely to be limited to niche applications where speed is the dominant requirement. The RSFQ technology was not included in the tables of the 2005 ITRS roadmap because it was claimed to be already in production and also because its applications are not in line with those targeted by CMOS devices.

Superconducting circuits of Josephson junctions can also be used for quantum information processing. A superconducting loop of three Josephson junctions has been proposed and demonstrated as a quantum bit or qubit [93, 94]. A coherent superposition of two persistent current states can be obtained when the two classical states are coupled via quantum tunneling through an energy barrier. The classical states of persistent currents can also be used as two binary bits [95]. Logic functions can be realized by coupling two or more bits, i.e., the circuit loops [96]. The interaction between loops is performed via magnetic interference of the superconductors.

The magnetically sensitive transistor, which is also known as spin transistor, is a hybrid magnetic/semiconductor transistor in which a magnetically controllable barrier is provided between a semiconductor base and collector to control the diffusion of charge carriers to the collector [97–100]. With the spin transistor, the charge carrier populations are distinguished by the direction of the spin or magnetic moment of the carriers instead of the electronic charge. A spin injector is used to spin-polarize the charge carrier population, so that the population has a selected magnetic moment. This population may or may not be enabled to flow to the collector via the magnetic barrier. The spin of an electron is semi-permanent and can be used as a mean of creating cost-effective non-volatile solid-state storage that does not require the constant application of current to maintain its state. It is one of the technologies being explored for the development of magnetic random access memory (MRAM) [101–103]. Spin transistors can also be used to synthesize NAND/NOR, AND/OR reconfigurable gates [104].

3.7 Overview of Nanodevices' Characteristics

The research on new and emerging architectures follows the same objectives that drives the research of novel devices. New computing models have been proposed to take advantage of the characteristics of emerging devices or to explore the use of CMOS devices in specific applications. Some existing or proposed electronic devices which potentially could reach the nanoscale are presented in Fig. 3.8 along with their current development status and most important related problems. The research and development status of these devices significantly varies, but this does not reflect the potential of a certain nanodevice to become a viable replacement or complement to CMOS technology; it merely reflects the interest shown by academia and industry up to date.

Figure 3.9 shows a comparison of technology densities that can be expected with each emerging logic device circuit and Fig. 3.10 shows the speeds that are projected

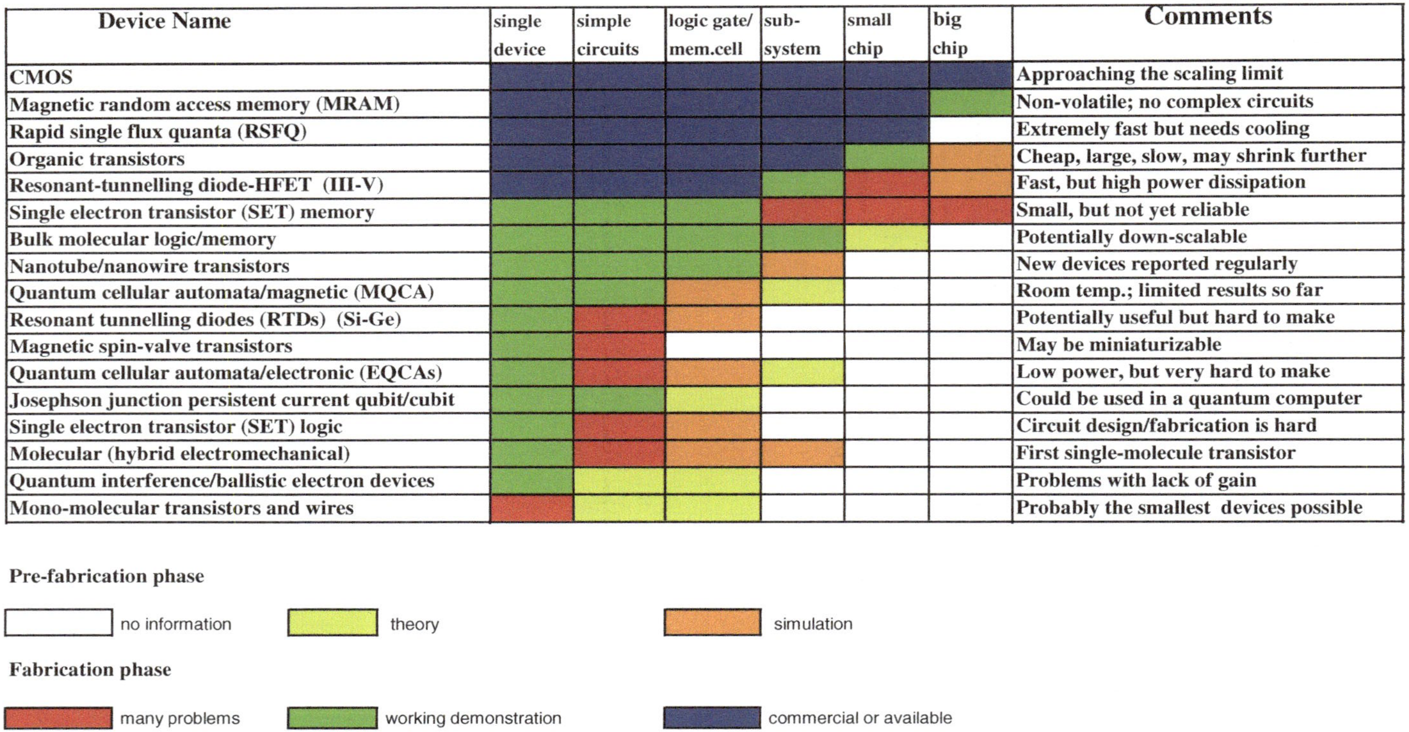

Fig. 3.8 Some existing or proposed electronic devices, which potentially could reach the nanoscale. The column headings "single device" … "big chip" are only intended as a means of ranking the degree to which large-scale integration has been achieved, which is a crude measure of their architectural complexity (adapted from [105])

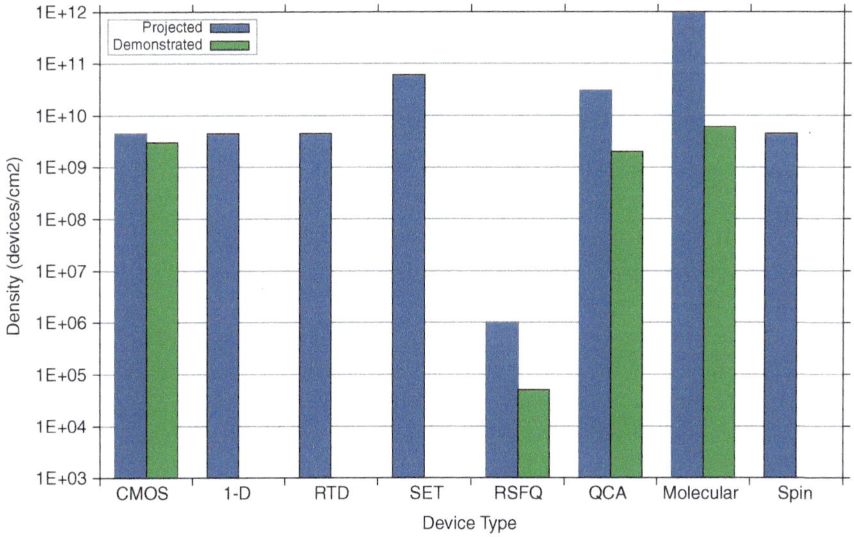

Fig. 3.9 Density (devices/cm^2) of CMOS and emerging logic devices (after [61])

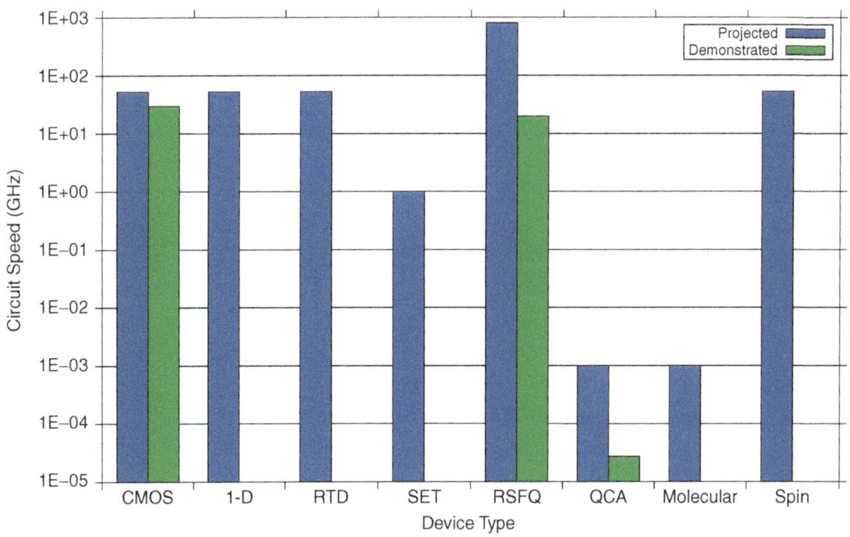

Fig. 3.10 Circuit speed (GHz) according to devices implemented (after [61])

to be achieved with these circuits [61]. The presented values reflect circuit operation and not individual device characteristics. Only considering the density and speed prospects, it is possible to see that the density advantage of some technologies does not reflect into speed advantages and vice versa.

3.8 Challenges for Designing System Architectures Based on Nanoelectronic Devices

Research on future nanoelectronic devices and architectures faces different challenges. On the one side, new devices are experimentally demonstrated, but there is a strong demand of accurate modeling of their behavior to enable the design of increasingly complex systems. On the other side, new architectures are proposed, described, and simulated, and there is a necessity to demonstrate that their manufacturing is possible in large scale with the expected accuracy. Research work on future architectures and devices is based on assumptions that are waiting to be proved. Many published works only focus on some aspects of the design space, making direct comparisons difficult. Reconfigurable crossbar architectures, for example, are the most researched alternatives so far, due to their claimed advantages, namely regular self-assembled low-cost fabrics, high integration density, low-power operation, and defect-tolerant capabilities. Self-assembly methods have already been demonstrated, but they are in an early stage of development, and it is assumed that they will enable generating regular full-sized structures, creating components with the desired properties. Although a low manufacturing cost is expected, at this point this is speculative, because of immature fabrication methods. Even with a tuned manufacturing process, research indicates that CMOS and nanostructures will have to be integrated together, and manufacturing costs will not be lower than today's CMOS-based systems. The nanofabrics concept trades simpler manufacturing by an increased complexity in post-fabrication procedures, and thus, low manufacturing costs may not mean low-cost chips.

The work of Stan et al. [86] presents a scenario of the challenges related to the electronic evolution, discussing all aspects from devices to architectures. Starting from the reasons for bottom-up assembly paradigms, the work focuses on the alternatives to CMOS evolution, the natural choice of crossbar arrays and mesh structures, the problems associated with them, the need for defect tolerance, and the integration of CMOS and nano(structures), called nano on CMOS, to guarantee scaling. As referenced in almost all of the works in the area, one of the characteristics that must be taken into account when evaluating an architecture is its capacity to tolerate permanent or transient errors that will be present at higher rates in future technologies [2].

Reviewing some major characteristic aspects that are likely to influence system-level design is relevant to identifying the important challenges involved in the design of system architectures based on nanoelectronic devices. These important aspects and issues include [42]

- reliability REL: reliability must be increased through redundancy in space, or time, or both, but the redundancy factors should be (very) small;
- testing TST: testing-associated costs have to be reduced;
- power-heat P/H; power and heat dissipation must be reduced or limited; this also includes power delivery and distribution, heat removal, and dealing with hot spots;
- connectivity CONN: connectivity has to be reduced both as overall wire length and as number of connections;
- hybrid integration HYB: hybrid integration must be achieved in the near term, including mixed design and interfacing;
- logic and (en)coding L/C: logic and coding must be optimized to reduce switching, computations, and communications (e.g., non-Boolean, error correction, spikes).

Other challenges, which may be considered, are algorithmic improvements ALG (e.g., probabilistic) and reduced design complexity DCOM (e.g., by applying design reuse). After a detailed analysis presented in [42], two factors emerge as the most influential, namely (i) reliability and (ii) power heat, with reliability appearing as the most important factor. Since power is already established as an important factor in the design of CMOS systems, it will continue to be one of the key factors in nanoelectronic systems. However, reliability is gaining importance and is assumed to be the fourth optimization pillar of nanoelectronics, along with the well-known triplet power/area/speed [16].

Unfortunately, reliability problems for technologies beyond CMOS are expected to increase significantly. The introduction of new materials could sharply decrease reliability margins. Beyond CMOS, device failure rates are predicted to be as high as 10% (e.g., background charge for SET [7]) increasing to 30% (e.g., self-assembled DNA [8, 9]). As a recent example, Green et al. [10] have reported defect rates of 60% in a 160-kb molecular electronic memory. Clearly, achieving 100% correctness at the system level using such devices and interconnects will not only be outrageously expensive but may be plainly impossible. Hence, relaxing the requirement of the 100% correctness for devices and interconnects may reduce the costs of manufacturing, verification, and test [15]. Still, this will lead to more transient and permanent failures of signals, logic values, devices, and interconnects. These conflicting trends will render technologists unable to meet failure rate targets and impose the delegation of reliability qualification to designers, i.e., failures will have to be compensated at the architectural level [106].

Regarding fault types, permanent errors will be dominating in nanodevices, mainly due to problems in fabrication, alignment, self-assembly, etc., and partially due to background charge fluctuations. However, intermittent and transient errors will also be present due to noise requirements and other sensitivities of nanodevices. Therefore, any fault-tolerant measure needs to cover all error types with special emphasis on permanent errors.

Accurate fault modeling, both at device and gate levels, is essential for successful reliability estimation. Currently, almost no models exist that can be used to precisely estimate the manufacturing defects or transient error rates in future nanodevices [105, 107, 108]. This enhances the need to develop accurate fault models. Some recent activities in this domain are presented in [31, 109]. In this book, a simple fault model of the SET is developed and is used in the simulations in Section 6.4. Moreover, a general framework that is used for fault modeling and reliability evaluation, and that can also be applied to nanodevices, is developed in Chapter 7.

The global picture is that reliability appears as one of the greatest threats to the design of future integrated computing systems. For emerging nanodevices and their associated interconnects, the expected higher probabilities of failures, as well as the higher sensitivities to noise and variations, could make future chips prohibitively unreliable. The result is that the current IC design approach based on the conventional zero-defect foundation might simply not be appropriate. Therefore, fault- and defect-tolerant techniques that allow the system to recover from manufacturing and operational errors will have to be considered from the (very) early design phases.

In the following chapter, we are giving an overview of existing reliability and fault-tolerant concepts in order to research their potential in nanotechnology applications.

Chapter 4
Fault-Tolerant Architectures and Approaches

Ever since humans first fashioned tools, they have had to ponder on their reliability and cope with the consequences of their failure. The unprecedented complexity of electronic appliances in the digital age has fostered the study and practice of fault tolerance, with the objective of delivering acceptable performance, even during sub-optimal or adverse circumstances. Over the past 50 years, fault tolerance has steadily advanced in stride with the permeation of computers into all aspects of society and human welfare [42].

The study of fault tolerance as we know it today emerged from the science of information theory (or informatics). In the course of a decade, Claude Shannon (1948), Richard Hamming (1950), Edward F. Moore (1956), and John von Neumann (1956) developed the fundamental principles of error correction and redundancy. These basic principles were immediately put into practice by the fledgling telecommunications, computing, and avionics industries in need of reliability. William H. Pierce (1965) unified the theories of masking redundancy, and shortly thereafter Algirdas A. Aviÿienis (1967) integrated these techniques for detection, diagnosis, and recovery into the concept of fault-tolerant systems. In the future, larger number of devices will be deployed in many applications and embedded systems (tera-scale integration), and reliability could turn out to be a showstopper for economically viable technology scaling. Thus, there is very high pressure to make sure that future nanoelectronic systems will be functioning correctly, over their expected lifetime, and even if they are not free of faults and defects [42]!

The methods presented in the literature are most commonly designed for, or demonstrated with single error. As technology progresses toward nanoscale devices, the defect density is expected to increase and the scenario involving multiple errors will be faced. Therefore, methods capable of tolerating several failures are the main focus of this chapter.

The well-known approach for developing fault-tolerant architectures in the face of uncertainties (both permanent and transient faults) consists of incorporating redundancy [13]. Redundancy can be either static (in space, time, or information) or dynamic (requiring fault detection, location, containment, and recovery). The word static refers to the fact that fault tolerance is built into the system structure and it efficiently masks the fault effects. The effect of dynamic redundancy is based

M. Stanisavljević et al., *Reliability of Nanoscale Circuits and Systems*,

DOI 10.1007/978-1-4419-6217-1_4,

on active actions as opposite to the passive operation of static redundancy. In this chapter, a detailed overview of static redundancy techniques is presented. A review of dynamic redundancy technique (reconfiguration) is provided in the end, in Section 4.2 as well.

4.1 Static Redundancy

Static redundancy can be categorized into space (hardware), information, and time redundancy, according to the resource that is used to create the redundancy. Also a combination of these can be used representing hybrid redundancy.

4.1.1 Hardware Redundancy

Hardware redundancy generally means replicating the functional processing module and providing a voting circuit to decide the correct output value based on redundant module outputs. Hardware redundant architectures mitigate the effects of faults in the devices and interconnects that make up the architecture and guarantee a given level of reliability. Higher reliability is gained because when a redundant component fails, the voter can decide the correct output based on the results of other redundant modules. The basic principle can be used at many different abstraction levels; the modules can be not only as simple as single gates but also as complex as whole processors or even larger constructions. The voter can be a simple bitwise hardware implementation or software algorithm running on a processor. General to all hardware redundancy realizations is the need for extra space or chip area. Thus, the methodology is also called physical, area, structural, or space redundancy [26].

Space (hardware) redundancy relies on voters (e.g., generic, inexact, midvalue, median, weighted average, analog, hybrid) and includes among others the well-known modular redundancy, cascaded modular redundancy, interwoven redundancy, and multiplexing schemes. Some other recently proposed techniques exist which are also discussed in the following.

4.1.1.1 *R*-Fold Modular Redundancy (RMR)

The concept of RMR (also known as N-tuple modular redundancy, NMR) consists of R functionally identical units working in parallel and comparing their outputs using a voter to produce the final output (see Fig. 4.1a) [13, 110, 111]. The units can be gates, logic blocks, logic functions, or functional units. Therefore, this technique can be used at many different levels of the design hierarchy.

The most common hardware redundancy realization is the triple modular redundancy (TMR), which consists of three redundant modules and a voting circuit. The voter normally performs majority voting (MAJ), which means that the output is the same as the output of two out of three consensual modules. TMR is capable of masking the output of a single failing processing module. The weak point of this realization is the voting circuitry, since a fault in voter could cause the whole circuit to fail. This can be mitigated by also replicating the voter and connecting the module outputs to all voters. This configuration is known as distributed voting RMR (see Fig. 4.1b) [112, 113] and will be a subject of an extensive study in Section 8.3.

4.1.1.2 Cascaded *R*-Fold Modular Redundancy (CRMR)

This concept (Fig. 4.1c) is similar to RMR, wherein the units working in parallel are RMR whose outputs are compared using the voter [6, 114, 115]. This configuration forms a "first order" CRMR. RMR can be considered to be "zeroth order" CRMR. Any order of cascading can be considered; however, the reliability of the final system does not necessarily increase with the cascading order number. Similar to RMR, the most common realization of CRMR is cascaded triple modular redundancy (CTMR).

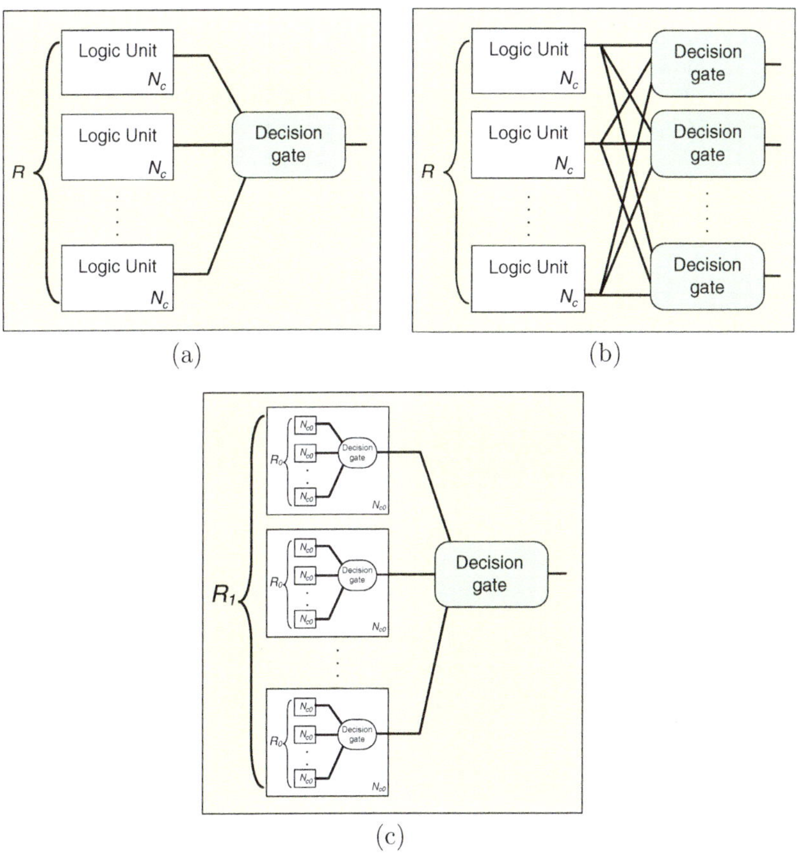

Fig. 4.1 **(a)** RMR; **(b)** distributed voting RMR; and **(c)** CRMR

Due to the area and latency overheads associated with this technique, the replicated units in CRMR with a multi-layer voting scheme are normally functional units or logic blocks, not single gates. Since the replicated functional units or logic blocks may consist of a large number of gates, their failure probability is higher than individual gates. Hence, the multi-level CRMR as shown in Fig. 4.1c may be used to partition the system into optimal-sized functional units or logic blocks to effectively allow the architecture to withstand an increased number of errors across the replicated units [116]. Optimal partitioning represents very important topic and is extensively explored in Section 8.3.

4.1.1.3 *R*-Fold Interwoven Redundancy (RIR)

The idea of RIR is based on interwoven redundant logic [117–119]. An RIR architectural configuration has *R* times as many gates and interconnects as the non-redundant network. The interconnections are arranged in random patterns. Such inherent randomness in the interconnections makes this structural redundancy technique favorable for the integration of molecular devices, since the manufacturing method for such devices is most likely to be based on stochastic chemical assembly. Figure 4.2 shows a non-redundant half adder and its corresponding triple interwoven redundancy (TIR) implementation. For a particular interconnect pattern, Han and Jonker [118] show that the RIR actually works as an RMR configuration, implying

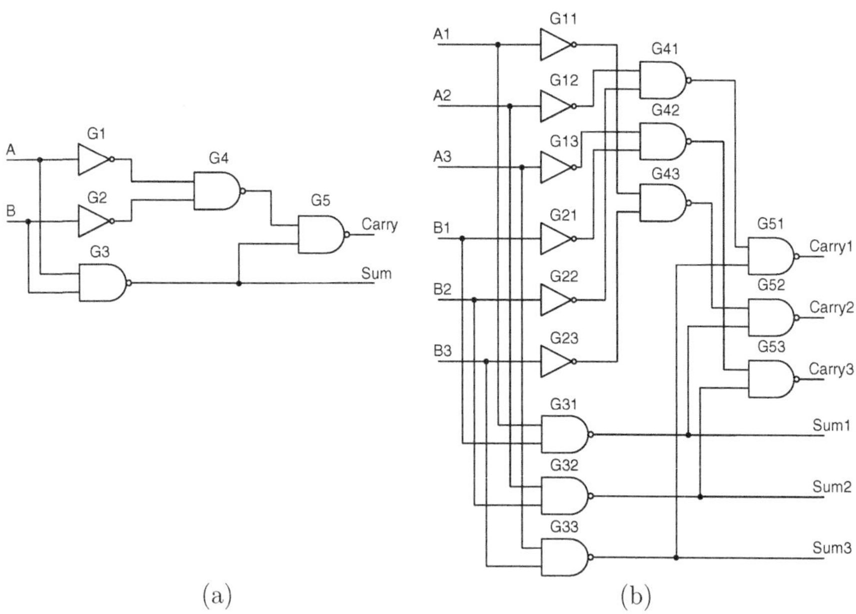

Fig. 4.2 A complementary half adder implemented with NAND logic: **(a)** non-redundant realization and **(b)** triple interwoven redundancy

that RMR is a specific implementation of RIR. Quadded logic [119, 120] is an ad hoc configuration of the interwoven redundant logic. It requires four times as many circuits, interconnected in a systematic way, and it corrects errors and performs the desired computation at the same time.

4.1.1.4 Multiplexing Techniques

Structural redundancy-based architectures that can circumvent transient faults can affect both the computation and communication in nanosystems. Interestingly, von Neumann addressed this issue in 1956 and developed a technique called multiplexing, trying to solve the problem of constructing a reliable system (automata) out of unreliable components [13]. He introduced multiplexing as a technique for constructing a system whose malfunction cannot be caused by the failure of a single device or a small set of devices. It has been identified as one of the most effective techniques for transient fault mitigation. Von Neumann proposed multiplexing architectures based on two universal logic functions – NAND and MAJ (majority).

In essence, the basic technique of multiplexing is similar to RMR, but instead of having a majority gate to decide on the proper output, the output is carried on a bundle of wires, e.g., for a single bit output one would have R wires (or N_{bundle} if we use von Neumann's notation) in a bundle which carries the output to the next stage. In this method, processing units of any size are replaced by multiplexed units containing N_{bundle} number of lines for every single input and output. Essentially, a multiplex unit consists of two stages. The first, the executive stage, performs the basic function of the processing unit in parallel. The second, the restorative stage, reduces the degradation caused by the executive stage and thus acts as a nonlinear "amplifier" of the output; see Fig. 4.3. An example of the executive stage given in Fig. 4.3 is a simple NAND (2-input gate); but it could be a unit with an arbitrary number of gates.

Besides original multiplexing introduced by von Neumann [13], techniques such as enhanced von Neumann multiplexing [121, 122] and parallel restitution [123]

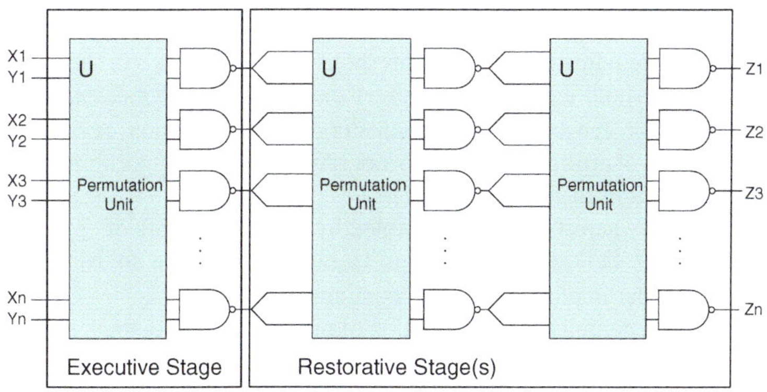

Fig. 4.3 NAND multiplexer

have been presented. A significant work has been done in this area for the past 50 years. Recent results in [31, 55, 124–126] are very promising.

4.1.1.5 Voters

The importance of voters in hardware redundancy techniques has been proven [26]. In recent works of Nikolic et al. [6, 127] and Stanisavljević et al. [115, 128] the voter has been identified as the component that limits the performance of hardware redundancy techniques, the most.

The voting algorithms can be distinguished according to their functionality, as generic and hybrid voting algorithms and as purpose-built voters. Generic voters only use the information of input signals to produce the output, while hybrid voters also use extra information such as the reliability of different modules or history of previous voting. Generic voters create the output according to the present output values of redundant modules. The most common algorithm is the exact majority voting. This is easily achieved in bitwise voting because the only possible values are logic-0 and logic-1.

In inexact voting, the output is chosen from the region that contains the majority or plurality of outputs. The selection of the output value can be a random selection of one of the majority values or it can be mid-value selection (MVS), where the output is counted as the mid-value of the majority or plurality outputs [129, 130].

Median voting forms another voting scheme, where the median of all the module outputs is selected as the voter output. An efficient software realization consists of sorting output values and subsequently selecting the $[(n+1)/2]$th value as the output, where n is the (odd) number of redundant modules [131].

In the weighted average voting, every module output is assigned a weight and the output is processed as the average of the module outputs multiplied by the weights. The weights are adjusted to obtain a desired input to output transfer function. An example is a fuzzy voter, which uses fuzzy set theory to adjust the weights [132]. Circuit realization examples of voters include the weighted bit-wise voters with threshold used with self-purging systems [133, 134] and the analog-weighted average voter [135] together with threshold circuit using capacitive threshold logic (CTL) [136]. The adjustment of the threshold is a crucial task in the operation of the circuit. Threshold can be static, based on circuit realization, can be set after manufacturing, or can be dynamic, adjusting to the operation environment. For example, the use of artificial neural network (ANN) learning algorithms in adjusting the thresholds has been suggested [137].

Analog voter where voting is preformed by analog comparators [138, 139] and perceptrons [140, 141] which represents averaging voter with adjustable threshold are important voter implementation in existing technologies.

Hybrid voters combine the information of present module outputs and additional information related to the module circuits or output sequence. For example, in a voting procedure that is based on previous data, weights in the weighted average voting are adjusted according to previous output values [134].

Different voting schemes are more appropriate for some applications than the others. The important issue related to voters is the implementation, since in complex implementations the probability of voter failure is increasing.

A new technique based on averaging and adjustable thresholding voter forms a central focus of this book and is presented in Chapter 6. Results and detailed evaluation of performance of this novel fault-tolerant technique are given in Sections 6.2 and 8.1.

4.1.2 Time Redundancy

Time (or temporal) redundancy trades space for time (e.g., alternating logic, recomputing with shifted operands, recomputing with swapped operands), where the basic principle consists of using the same resource many times, and compares the results obtained from different rounds of computation. As opposed to hardware redundancy, time redundancy entails repetition of the same computation, if the computation fails to compute correctly. A classical temporal redundancy mechanism used in hardware is based on checkpoints and roll-back recovery. The method based on repeating the same computation many times is effective to detect transient errors, but permanent, and in many cases also intermittent errors occur at the same place during all calculations and therefore cannot be detected and corrected. This problem can be overcome by encoding the operand before processing and decoding afterward.

Time redundancy methods differ in the way encoding and decoding is accomplished. The simplest coding method is complementing, which is used in alternating logic. In order to be able to use this coding, the self-duality of the circuit is required or possibly extra input is needed.

The recomputing with shifted operands (RESO) uses the shifting of operands to the left prior to calculation and back to the right after calculation. This method demands extra width for operands or cyclic shift to be used, which on the other hand means complex logic for carry signals on adder circuits.

Another coding scheme is recomputing with swapped operands (RESWO), where the upper and lower parts of the operands are swapped before calculation and reversed back after it. The method needs no extra bits, and the logic for handling carry bits in adder circuits is more straightforward than in RESO [113]. The error-correcting properties are gained by repeating the operation at least three times and performing voting over the three results. Different coding is used for different calculation rounds, e.g., no coding, shift to the left, and shift to the right. Bit-wise majority voting can be problematic because the arithmetic operations commonly affect many bits [113].

4.1.3 Information Redundancy

Information redundancy, i.e., providing extra information is used in applications related to error detection and correction to make systems more reliable. Some basic information redundancy techniques are given in the following. The most common error detection methods are parity and cyclic redundancy check (CRC).

The parity technique uses a parity bit as an error detection code. The obvious drawback is that an error correction is not possible, as there is no way to determine which particular bit is corrupted. If an error occurred during transmission, then the entire data must be discarded and re-transmitted.

The CRC is a very powerful and easily implemented technique to obtain data reliability based on a hash function which is used to produce a checksum. The checksum holds redundant information about the block of data that helps the recipient detect errors. A CRC is computed and appended before transmission or storage and verified afterward by the recipient to confirm that no change occurred during transmission. It is one of the most widely used techniques for error detection in data communications. The technique is popular because CRCs have extremely efficient error detection capabilities, have little overhead, and are easy to implement. Moreover, they are simple to implement in binary hardware and are easy to analyze, mathematically. The parity technique represents a special case of CRC.

An error-correcting code (ECC) is an algorithm for expressing a data signal such that any errors which are introduced can be detected and corrected, within certain limitations, based on the other parts of the signal. As opposed to the parity and CRC, even the simplest ECCs can correct single-bit errors and detect double-bit errors. It is used in computer data storage and transmission. There are other codes which can detect or correct multi-bit errors.

The most common ECC is the Hamming code, which uses the concept of overlapping parity including multiple parity bits, and every data bit is covered with several of them. The modified Hamming code enabling both correcting single errors and detecting double errors can be achieved by adding one extra check bit, which is used as the parity bit of the whole code word. Other examples of ECCs are BCH code, Reed–Muller code, binary Golay code, and convolutional code [142]. ECC-based computer memory provides higher data accuracy and system uptime by protecting against soft errors [113].

4.1.4 Hybrid Approaches

Methods that combine the aspects of many different redundancy types are called hybrid approaches. The method which combines hardware redundancy or more specifically triple modular redundancy and time redundancy is called time-shared triple modular redundancy (TSTMR). Three identical processing elements and a voting circuit are present in this technique, as in TMR. The time domain approach is applied in a way where every operand is divided into three parts.

The procedure starts by performing the target operation on the lower part of the operands; the result is voted among the three module outputs and saved into a register. In the next step, the same operation is performed with the middle parts and finally with the upper parts of the operands. When all parts are calculated, the results are combined to create the final result. Special logic is inserted to handle the carry propagation from one phase to another. The benefit of the method is the lower area overhead than in TMR and also the time required for computation.

The same method is also called recomputing with triplication with voting (RETWV) [143], hardware partition in time redundancy (HPTR) [144], or recomputing with partitioning and voting (RWPV) [145] and its usage is presented for adders and multipliers [143] as well as for dividers [146]. An extension of the same methodology is quadruple time redundancy (QTR), where operands are divided into four parts and the computation has four phases [147, 148].

4.1.5 Recent Techniques

Some fault-tolerant techniques, recently being proposed, even though not necessarily implementing redundancy in a classical sense, are in the first place hardware techniques. These techniques can more appropriately be categorized into fault avoidance instead of fault tolerance, since they tend to reduce the occurrence of faults through a reduction in their severity of impact, or so-called single-event upsets (SEUs). The main source of SEUs is cosmic radiation and energetic particles.

The gate hardening technique [149–152] is based on identifying the weakest gates by reliability evaluation and improving their reliability by simultaneously applying dual-VDD and gate sizing techniques. The gate hardening technique is appropriate for single-event upsets, but it can be modified to be applied to permanent errors.

Partial error masking is also designed for SEUs and should improve reliability by identifying the weakest gates by reliability evaluation and apply TMR only to those gates [153]. The dominant value reduction technique [153] is also applied.

Concurrent error detection (CED) is based on the synthesis of approximate logic circuits as a low overhead, non-intrusive solution to enhance reliability combined with error masking [154, 155].

4.2 Dynamic Redundancy

As opposed to static redundancy, redundant parts of the design in dynamic redundancy systems are only activated, dynamically upon the need to counteract the appearance of a fault. The use of dynamic redundancy means the introduction of special control circuitry and elements. The design of these control parts may turn out to be intricate. The benefit gained with dynamic redundancy is better reliability, especially in the occurrence of permanent and multiple errors, and quite often reliability is also gained with smaller area overhead than in the corresponding static redundancy approach. However, dynamic redundancy is not suitable for transient and intermittent errors, since the time necessary to detect a fault and activate a redundant circuitry often largely surpasses the time limit to suppress the impact of such an error.

In principle, the dynamic redundancy operation can be divided into four phases, namely fault detection, fault location, fault containment, and fault recovery. After the detection of a fault, a fault needs to be located, the error source isolated, and in the final phase fault recovery usually represents reconfiguration of the circuit so that the erroneous part is disabled [113]. One major approach including dynamic redundancy is reconfiguration.

4.2.1 Reconfiguration

In 1998, a paper was published in SCIENCE on the "Teramac" reconfigurable computer [5] (Fig. 4.4), supporting the proposal that this technique would be useful for overcoming manufacturing defects in nanocomputers. The Teramac experiment was one of the first reprogrammable computer architectures implemented as a bottom-up assembly of basic components. Designed to be a custom-configurable computer (CCC), the Teramac was built with field-programmable gate array (FPGA) chips that were responsible for logic operations and redundant interconnections in the form of crossbars or fat tree networks. From the beginning of the project, the Teramac was

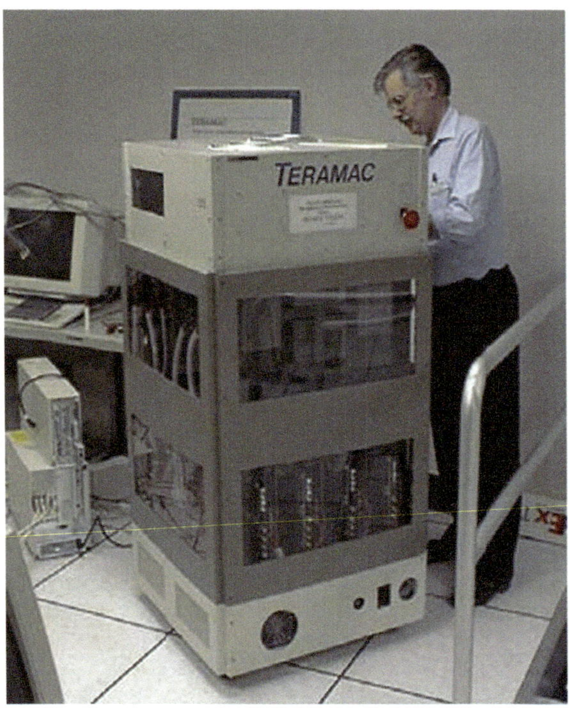

Fig. 4.4 Teramac, with David Kuekes, one of its architects (adapted from [156], © [1998] IEEE)

designed to be a defect-tolerant architecture, and its implementation was based on unreliable components. The Teramac used 864 FPGAs in its structure, and 647 of them had some kind of defect. A total of 3% of all resources in the architecture was defective, but the associated problems were circumvented by the extremely high degree of interconnections implemented. Test procedures and defect mapping were driven by an independent workstation, but after determining a small reliable portion of the structure, the Teramac could be programmed to test itself. The Teramac computer showed that it is possible to build defect-tolerant nanoarchitectures using only wires, switches (the crossbar), and memory (the look-up tables in the FPGAs). Its architecture allowed highly parallelized computing, and the Teramac could achieve high performance with a low operation frequency.

The work of Lach et al. [157] provided a theoretical basis for Teramac experiment. In this work, devices are assembled in programmable logic elements (FPGA), like a configurable logic block (CLB, shown as a sub-unit in Fig. 4.5), and interconnects which can be configured to implement any logic circuit. A number of these CLBs are later grouped together in so-called atomic fault-tolerant block (AFTB, larger units in Fig. 4.5). It is assumed that the AFTB can be configured to perform some basic set of operations, even though any one of its constituent CLBs may be faulty. In general, different types of AFTBs can be designed to carry out different functions, and each type may incorporate different numbers of CLBs.

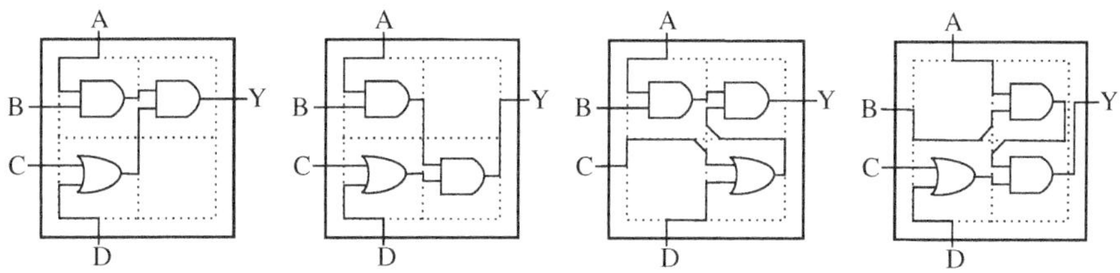

Fig. 4.5 The basic structure of the reconfiguration technique theory (adapted from [157], © [1998] IEEE)

It is expected that reconfigurable fabrics made from next-generation fabrication processes will undergo through a post-fabrication defect mapping phase during which these fabrics are configured for self-diagnosis [5, 158, 159]. Thus, fault tolerance in such fabrics can be achieved by detecting faulty components during an initial defect mapping phase and excluding them during actual configuration. Thus, the design of reliable digital logic and architectures using unreliable nanofabrics will require defect mapping followed by defect avoidance to circumvent hard errors. Defect mapping is the process of finding defect locations in a nanofabric, and defect avoidance is the process of mapping a computing logic on a faulty nonofabric, knowing its defect maps. While such reconfigurable architectures may aid in circumventing manufacturing defects at the nanoscale, they will not provide tolerance to natural external transient errors. The addition of structural redundancy a priori may enhance the reliability of such systems in the presence of transient errors.

Even though reconfiguration is reported to be the most effective technique to cope with manufacturing defects in nanodevices [6, 160], some serious drawbacks remain. Besides lacking the possibility to tolerate transient errors, reconfiguration needs long testing and configuration time, especially considering the prospects of building trillion-transistor logic systems with high defect densities [121, 127]. Testing and configuration time in such a large system where 10% or more devices are defective can reveal itself an intractable task.

Another issue is that fault-tolerant techniques such as system reconfiguration have a cost in terms of connectivity and control overhead. Reconfiguration methods usually rely on FPGA-based architectures. Present-day FPGA chips only have one-tenth of the area dedicated to effective logic. For example, current microprocessor chips contain $\sim 10^8$ devices, whereas FPGA only chips contain $\sim 10^7$ effective devices, the remainder of the chip being occupied with wiring, routing, and control elements. If the same architectural concepts are carried over into future nanoelectronic systems (as is explicitly acknowledged in [161], for example), then future nanoelectronic computers will always fall short from the optimum performance by a factor of at least 10, in spite of how energy efficient their constituting devices are.

4.3 Overview of the Presented Fault-Tolerant Techniques

The feasibility of designing reliable nano architectures using practical or economical (i.e., small and very small) redundancy factors is an important challenge related to fault-tolerant architectures. Therefore, the main criterion when evaluating different topologies is an overhead represented by the redundancy factor. Other important criteria include the ability to tolerate any error type and the difficulty to implement the given fault-tolerant techniques into future designs including 10^{12} devices.

Regarding permanent errors which tend to be the dominant type of errors in future nanotechnologies, only static redundancy techniques offer a viable solution. Reconfiguration, even though having excellent performance regarding necessary redundancy to cope with high defect densities, has to be enhanced with static redundancy in order to mitigate intermittent and transient errors.

Regarding static redundancy, techniques like classic RMR and CRMR offer the optimal performance for relatively low redundancy factors, but can also tolerate only low device failure probability (10^{-8}–10^{-7}) [6, 114, 115]. On the other hand, techniques such as NAND/MAJ multiplexing can tolerate much higher device failure probability (10^{-3}–10^{-2}), requiring moderate redundancy factors (10–100) [31, 121–123, 126, 162].

One solution to enhance the performance of RMR and CRMR techniques consists of including better voters and to use distributed voting. This direction of research comprises an important part of this book which is presented in Chapter 8.

Not all the fault-tolerant techniques are applicable at all the design hierarchical levels. Some techniques are more suitable for circuit level (level of logic gates), others for high level (processing cores or chips), and some have a more universal applicability. However, the applicability level is an important property of fault-tolerant techniques, and as such takes a significant place in the study presented in this book, mainly in Chapter 8. As an illustration of applicability of fault-tolerant techniques, a brief overview is depicted in Fig. 4.6.

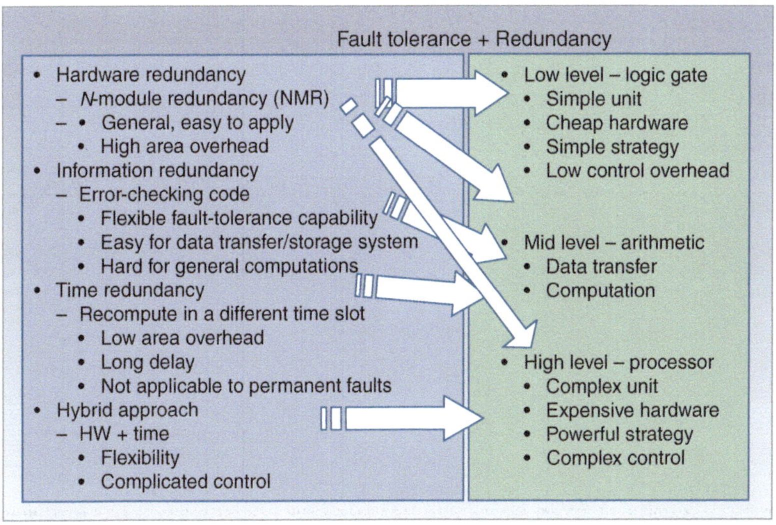

Fig. 4.6 Fault-tolerant approaches and their applicability at various levels (adapted from [163], © [2007] IEEE)

Chapter 5
Reliability Evaluation Techniques

The expectations are that the future nanocircuits will exhibit higher frequency of failures. The higher density of transistors on-chip is one of the reasons for this behavior. In particular, as feature sizes are aggressively scaled, the processing of ICs becomes more complex and inevitably introduces more defects. Other factors such as geometric variations, or related to the tiny amounts of energy which are required to enable the switching of nanodevices, make them susceptible to transient failures and negatively impact on reliability. Architectures built from emerging nanodevices will be extremely susceptible to parameter variations, fabrication defects, and transient failures induced by environmental/external causes [2, 164].

Therefore, the design community has been urging computer aided design (CAD) researchers to pay more attention to reliability issues. This was the message from the ICCAD'06 conference, which gathered both communities (design and CAD) in San Jose, in November 2006. The current general state-of-the-art electronic design automation (EDA) tools take only delay, power, and area as important optimization parameters. However, the increased importance of reliability strongly suggests that reliability should be included as an additional fourth optimization parameter in the forthcoming EDA tools. In order to establish reliability as the fourth optimization parameter, the tools for accurate reliability evaluation are necessary. An overview of existing reliability evaluation tools is provided in this chapter. The chapter is organized as follows. In Section 5.1 the tools that have left an important mark on the course of reliability evaluation development are presented. The most recent tools are presented in Section 5.2. Finally, a Monte Carlo tool developed by the authors, including a detailed description of the algorithm and its realization, is given in Section 5.3.

The methods to determine the reliability of logic circuits can generally be divided into two groups:

- analytical and
- simulation based.

A model is an abstraction of the various assumptions about a system's behavior. These assumptions represent mathematical or logical relationships, which, if they are simple enough, lead to analytic solutions [165]. *Analytical* evaluation

M. Stanisavljević et al., *Reliability of Nanoscale Circuits and Systems*,
DOI 10.1007/978-1-4419-6217-1_5, © Springer Science+Business Media, LLC 2011

can be applied for small circuits or in regular topologies without losing accuracy. Analytical evaluation can also be applied to perform an approximate evaluation of different fault-tolerant topologies (as in the seminal work of von Neumann [13] and the works of Siewiorek and Swarz [110], Depledge [111], and Spagocci and Fountain [114]), as well as to perform the evaluation of reliability bounds of any type of circuits built with certain gates (as in the early works of Dobrushin and Ortyukov [166, 167]). Similar analyses have been performed in more recent works [168–175]. Fault-tolerant topologies have been analytically studied in recent works of Nikolic et al. [6, 127] and Stanisavljević et al. [115, 128]. Numerous analytical studies related to the use of multiplexing architectures have been published [55, 121–123, 126]. An interesting approach in the analytical evaluation of reliability through noise and parameter variation modeling is presented in [176–179].

The detailed analytical evaluation of redundant fault-tolerant architectures is conducted in Section 8.3.

However, as more details of reality are introduced into the models, analytical solutions become intractable and *simulation* emerges as a reasonable method to determine the operational characteristics of the model. Simulation involves numerically evaluating a system over some relevant period of time and using the data gathered to characterize the model's behavior. The methods used for simulating stochastic systems can be divided into

- experimental and
- numerical.

Experimental methods rely on implicitly performing the analysis by observing the results obtained from many experiment runs. The most popular experimental method is the discrete-event simulation (DES), which reproduces the behavior of the system. In order to accomplish this, DES relies on random number generators that sample the random activities of the analyzed system. Once the model is built, the computer performs as many sample runs from the model as necessary to draw meaningful conclusions about the model's behavior. Consequently, the DES analysis is indirectly conducted, based on the observation of many sample runs. The most prominent advantages of DES are its intuitiveness and its ability of simulating models for which deterministic solutions are intractable. The most popular experimental method is Monte Carlo simulation (MC), which mimics the behavior of the real system with parametric adaptation in each run. The MC method is by far the most widespread one in the semiconductor community [180]. It will be used in the future for analyzing the behavior of (novel) devices and gates, as well as small sub-circuits. However, MC is a very time-consuming process which limits its application in the nanoera mainly to simulations at the device, gate, and small block levels; nevertheless, the reliability results obtained could be stored as parameters of future libraries (of devices and gates).

A custom-built MC simulator incorporating specific failure models is used throughout this book. Its development and properties are described in Section 5.3.

Numerical methods are designed for analyzing stochastic models without incorporating any random behavior. The simulation results that they deliver are always the same for the same model parameters. These methods work by describing the flow of probabilities within the system, usually using differential equations and numerical methods for solving them. In the remaining of this section a brief chronological overview of existing tools incorporating numerical methods is presented considering two groups: (i) historically important tools and (ii) the most recent tools.

5.1 Historically Important Tools

A historical overview of reliability evaluation tools which are based on numerical methods is presented in the following. The interested reader can find many earlier results including REL70, RELCOMP, CARE, CARSRA, CAST, CAREIII, ARIES-82, SAVE, MARK, GRAMP, SURF, SURE, SUPER, ASSIST, SPADE, METASAN, METFAC, ARM, and SUPER, in the extensive review of Johnson and Malek [165].

The Hybrid Automated Reliability Predictor (HARP) tool was pioneered in 1981 at Duke and Clemson Universities [181]. HARP uses a fault tree analysis technique for describing the failure behavior of complex technical systems. Fault tree diagrams are logical block diagrams that display the state of a system in terms of its components. The basic elements of the fault tree are usually failures of different components of one system. The combination of these failures determines the failure of the system as a whole. Further development has led to the Symbolic Hierarchical Automated Reliability and Performance Evaluator (SHARPE) [182] (Duke University) and the Monte Carlo Integrated HARP (MCI-HARP) [183] (developed at Northeastern University).

In the early 1990s a few other tools providing numerical analyses have been developed: *TimeNET* at the Technical University of Berlin [184], *UltraSUN* (and later *Möbius*) at the University of Illinois at Urbana-Champaign [185], and *SMART* at the University of California at Riverside [186]. These were followed in the mid-1990s by the Dynamic Innovative Fault Tree (DIFTree) [187] and Galileo [188], both from the University of Virginia. Galileo extended the earlier work on HARP, MCI-HARP, and DIFTree using a combination of binary decision diagrams (BDD) and Markov methods and is currently commercialized by Exelix [189].

Probabilistic model checking (PMC) is an algorithmic procedure applied to confirm whether a given probabilistic system satisfies probabilistic specifications such as "the probability of logical correctness at the output of a logic network must be at least 90%, given that each gate has a failure probability of 0.001." The system is usually modeled as a state transition system with probability values attached to the transitions. Markov chains can be used for describing and analyzing models exclusively containing exponentially distributed state changes and are implemented in some of the reliability evaluation tools. Depending on the nature of the time domain, discrete-time Markov chains (DTMCs), continuous-time Markov chains (CTMCs), and Markov decision processes (MDPs) are applied. However, due to the usage of random number generators, these numerical methods sometimes fail to capture the real behavior of the simulated process [190].

In 1999 a team from the University of Birmingham introduced the Probabilistic Symbolic Model Checker (PRISM) [191]. PRISM relies on a probabilistic model checking for determining whether a given probabilistic system satisfies given probabilistic specifications. It applies algorithmic techniques to analyze the state space and calculate performance measures associated with the probabilistic model. PRISM supports the analysis of DTMCs, CTMCs, and Markov decision processes (MDPs).

NANOPRISM [192] is a tool built on the probabilistic model checker PRISM, which was developed at Virginia Polytechnic University. It uses model checking techniques for calculating the probabilities of transient failures within devices and interconnections of nanoarchitectures. RAMP [193] is a reliability tracking tool applied for the analysis of lifetime reliability that has two implementations, namely 1.0 and 2.0, which are different in both their efficiency and the assumptions they impose on the analyzed models. RAMP 1.0 is simpler and can be applied to real hardware and used in simulators. RAMP 2.0 allows more complex models to be analyzed and uses the Monte Carlo method to run experiments. However, it cannot be applied to real hardware because of an excessive computational complexity.

The *proxel*-based method [194] is introduced as an alternative to Monte Carlo method for simulating discrete stochastic models. Proxel is the abbreviation of "probability element." It describes every probabilistic configuration of the model in a minimal and complete way. Each proxel carries enough information to generate its successor proxels, i.e., to probabilistically determine how the model will behave [195]. This transforms a non-Markovian model into a Markovian one. This approach analyzes models in a deterministic manner, avoiding the typical problems of Monte Carlo simulation (e.g., finding good quality pseudo-random number generators) and partial differential equations (PDEs, difficult to set up and solve). The underlying stochastic process is a discrete-time Markov chain (DTMC), which is constructed on-the-fly by inspecting all possible behaviors of the model.

Since soft (transient) errors are becoming an important concern in digital integrated circuits, another important class of tools that aspires to accurately evaluate soft errors rates (SER) need to be mentioned. It has long been known that many soft faults are masked and do not lead to observable circuit errors. Therefore, evaluators are needed to assess the impact of masking mechanisms on the soft error rate (SER) of a circuit. Further, deliberately increasing masking is key to low-SER designs. Hence, SER analysis can effectively guide and evaluate synthesis by accounting for relevant masking mechanisms.

Recent SER evaluators include SERA [196], FASER [197], SERD [198], and MARS-C [199] along with its sequential extension MARS-S [200]. These tools estimate the SER of a technology-mapped circuit by accounting for three masking mechanisms with varying levels of detail. The three masking mechanisms are as follows [201]:

- logic masking (the glitch occurs in a non-sensitized portion of the circuit);
- electrical masking (the glitch is attenuated and blocked by the electrical characteristics of CMOS gates); and
- temporal masking (the glitch occurs in a non-latching portion of the clock cycle).

Logic masking is accounted for by explicit enumeration of the input vector (or state) space in decision diagram-based methods [197, 199] or by fault simulation on specific vectors [196, 198]. Electrical masking is assessed using SPICE-based pre-characterization of the gate library. Timing masking is either approximated as a derating factor proportional to the latching time of flip-flops in the design [197] or based on timing analysis information [197]. In addition, MARS-S [200] uses Markov chain analysis and symbolic simulation to analyze SER in sequential circuits.

While these methods offer detailed analysis of SER, they can be difficult to use during logic design because they require complete information such as electrical characterization and timing analysis, which may be unavailable during logic design, and they use unscalable methods for logic masking analysis. Some tools [190, 197, 199] use algebraic decision diagrams (ADDs) (decision diagrams with multiple real-valued terminals) to completely enumerate input patterns and calculate pattern-dependent error probabilities for the logic masking analysis, which has exponential worst-case complexity. This use of ADDs in SER analysis is different from the use of BDDs in logic synthesis to represent Boolean functions. The latter is generally much more efficient. Other tools electrically simulate circuits vector by vector, which can slow down SER analysis and become a bottleneck in circuit optimization as well. The most recent techniques perform the so-called signature-based SER analysis where signatures, i.e., partial truth tables generated via bit-parallel functional simulation, are used during soft error analysis and logic synthesis [202].

5.2 Most Recent Progress in Reliability Evaluation

A significant progress has been achieved in reliability evaluation, in recent years.

The *probabilistic transfer matrices* (PTMs) framework was first presented in [203], but the underlying concept can be traced back to [204]. Given a circuit C with n inputs and m outputs, the PTM for C is a $2^n \times 2^m$ matrix M whose entries are $M(\mathbf{i}, \mathbf{j}) = \Pr(\text{outputs} = \mathbf{j} | \text{inputs} = \mathbf{i})$, where $\mathbf{i}$ and $\mathbf{j}$ are input and output vectors, respectively. For example, the PTM for a NAND gate with a failure probability of ε is given by

$$\mathrm{PTM}_{\mathrm{NAND}} = \begin{bmatrix} \varepsilon & 1-\varepsilon \\ \varepsilon & 1-\varepsilon \\ \varepsilon & 1-\varepsilon \\ 1-\varepsilon & \varepsilon \end{bmatrix} \tag{5.1}$$

and its output probability is given by $\begin{bmatrix} p_0 \; p_1 \end{bmatrix} = \begin{bmatrix} p_{00} \; p_{01} \; p_{10} \; p_{11} \end{bmatrix} \times \mathrm{PTM}_{\mathrm{NAND}}$. A PTM for a specific circuit is formulated by composition of the gate PTMs, the composition being dependent on the logic dependency of the circuit.

The PTMs can be used to evaluate the circuit overall reliability by combining the PTMs of elementary gates or sub-circuits [190]. It performs simultaneous computation over all possible input combinations and *calculates the exact probabilities of errors*. Besides this accuracy, another advantage resides in that it is trivial to have different probabilities of failures for the different gates (see [205]). PTMs store the probability of occurrence of every input–output vector pair for each level in the circuit to compute the probability of error at the output of the circuit. If the largest level in the circuit has n inputs and m outputs, the straightforward PTM representation requires $O(2^{n+m})$ memory space. This leads to massive matrix storage and manipulation overhead. Even with compaction of the matrices using ADDs, the high runtimes for benchmark circuits with 20–50 gates suggest their inapplicability to large circuits. The practical limit size of the circuits that can be simulated is approximately 16 input/output signals. Moreover, the circuit's overall probability of failure requires manual dividing of the circuit into several stages, generating the PTM for individual stages, and finally combining all the stages PTMs to create the circuit's overall PTM. This process is fairly simple for small circuits, but it becomes very intricate and error prone as the circuit size increases.

The *Bayesian network* numerical method (BN) is a powerful modeling tool especially applicable to problems involving uncertainty [206–209]. The relation between circuit signals and Markov random fields was presented in the context of probabilistic computations. The name BN comes from its reliance on Bayes' theorem as the basis for updating information. Bayes' theorem states that the conditional probability of a set of possible causes for a given observed event can be computed from the knowledge of the probability of each cause and the conditional probability of the outcome of each cause:

$$P(A|B) = \frac{P(B|A)P(A)}{P(B)}. \tag{5.2}$$

A Bayesian network encodes the joint probability distribution over a set of variables $\{X_1, \ldots, X_n\}$, where n is finite, and decomposes it into a product of conditional probability distributions over each variable given its parents in the graph. Nodes with no parents, use the variable probability. The joint probability distribution over $\{X_1, \ldots, X_n\}$ can be obtained by calculating the product of all of these prior and conditional probability distributions:

$$P(X_1, \ldots, X_n) = \prod_{i=1}^{n} P(X_i|PA(X_i)), \tag{5.3}$$

where $PA(X_1)$ denotes the probabilities of the parent nodes of node X_i. The conditional probability of output(s) given input signals determines how errors are propagated through a circuit. Using this theoretical model, it is possible to predict the probability of output error given the gate errors. The probability of error is *exact*, similarly as in PTMs. However, BNs suffer from similar problems as PTMs. A massive matrix storage and manipulation overhead is involved, due to large conditional probability tables that support Bayesian networks. Although this problem is mitigated in the Bayesian network approach for small circuits, manipulating Bayesian networks for large circuits is potentially intractable.

The *probabilistic gate model* (PGM) is another analytical approach for estimating reliabilities of circuits [210, 211] which entails the formulation of a PGM for each logic gate type. The procedure to obtain gate's PGM can be formulated as follows:

$$p(\text{out}) = \begin{bmatrix} p_i & 1 - p_i \end{bmatrix} \cdot \begin{bmatrix} 1 - \varepsilon \\ \varepsilon \end{bmatrix}, \tag{5.4}$$

where ε is the gate failure probability, $p(\text{out})$ is the probability of the output of the gate being stimulated, and p_i is the probability that a fault-free gate will produce logic-1 at its output. For instance, the PGM of the NAND gate is

$$p(\text{out}) = \begin{bmatrix} 1 - p(A)p(B) & p(A)p(B) \end{bmatrix} \begin{bmatrix} 1 - \varepsilon \\ \varepsilon \end{bmatrix}, \tag{5.5}$$

where $p(A)$ and $p(B)$ are the probabilities of inputs A and B being stimulated, since for a NAND gate $p_i = 1 - p(A)p(B)$. Hence, $p(\text{out})$ is the probability of the output of faulty NAND gate being at logic-1. This formulation can be iteratively applied to compute circuit reliability. This method can be used for *any type of gate and fault models*. The PGM method divides the circuit into many small modules (i.e., gates), and input/output signals are assumed to be statistically independent. Hence, an overall reliability of a circuit can be obtained by multiplying the individual reliabilities of each output. In a circuit that includes fanouts, signals are correlated; hence, PGMs lead to approximate results. For circuit of m inputs and n gates, the memory complexity is $O(n)$ and the computational complexity is $O(n \cdot 2^m)$, which still makes it unscalable.

The *single-pass reliability analysis tool* [212] presents an implementation of a fast, accurate, and scalable, novel algorithm for reliability analysis. At the core of this algorithm lies the observation that an error at the output of any gate is the cumulative effect of a local error component attributed to the probability of failure of the observed gate and a propagated error component attributed to the failure of gates in its fanin cone. In the algorithm, gates are topologically sorted and processed in a single pass from the inputs to the outputs. Topological sorting ensures that before a gate is processed, the effects of multiple gate failures in the fanin cone of the gate are computed and stored at the inputs of the gate. The cumulative effect of failures at the output of the gate is computed using the joint signal probability distribution of the gate's inputs, the propagated error probabilities from its fanin stored at its inputs, and the error probability of the gate. The effect of reconvergent fanout on error probabilities is addressed using pairwise correlation coefficients.

The algorithm is very fast since there are no matrix multiplications; only one single multiplication is performed for each gate in the circuit. Moreover, the required memory space is small and the accuracy is high thanks to an efficient way of handling of reconvergent fanouts.

The *signal probability reliability analysis* (SPRA) [213, 214] presents an algorithm to compute the cumulative effect of faults in the gates of a circuit, where a (transient) fault is modeled as a bit-flip error at the output of the faulty gate. The algorithm takes into account the reliability of circuit gates and the topological structure of the circuit to determine the probability of correctness of the signals. The computation of the cumulative effect of errors embeds the probability of occurrence of multiple simultaneous faults in the target circuit. The algorithm applies matrix multiplications for gates similarly as PTMs. However, instead of keeping the probability information of the whole circuit in one PTM, SPRA only keeps the information related to the probability of each signal independently.

The signal probability is defined with the matrix of four possible states that represent the probabilities of the signal being a correct logic-0, a correct logic-1, a faulty logic-0, or a faulty logic-1 [215]:

$$P_{2\times 2}(\text{signal}) = \begin{bmatrix} P(\text{signal} = \text{correct logic-0}) & P(\text{signal} = \text{faulty logic-1}) \\ P(\text{signal} = \text{faulty logic-0}) & P(\text{signal} = \text{correct logic-1}) \end{bmatrix}. \tag{5.6}$$

Three algorithms are proposed to mitigate the effect of reconvergent fanout on signal error probabilities. They differ by accuracy and speed.

SPRA is scalable and fast since only one matrix multiplication is performed for each gate in the circuit. Moreover, the required memory space is small and the accuracy is high thanks to an efficient algorithm handling the reconvergent fanouts.

The overview of most recent tools is reported in Table 5.1 where tools are compared with respect to speed, accuracy, memory, and scalability. Data presented in the table are compiled from all references mentioned in this section. Since reported results are not always directly comparable some fields in the table are descriptive.

Table 5.1 Expressions for input error components

	Speed-up factor	Accuracy (max rel. err. %)	Memory requirements	Scalability
PTMs	1	Exact	Very high	No
BN	>500	Exact	High	No
PGMs	>500	Low accuracy	Low	No
Single pass	>250,000	<10%	Low	Yes
SPRA	5000–500,000	5–30%	Low	Yes

The majority of the methods presented in this section are intended to handle fault models of transient errors. However, all of them can easily be modified to evaluate reliability with fault models encompassing permanent errors.

The main problem with most of the existing approaches relates to the fact that they assume single value of the probability for each gate (as customary done in the reliability literature) which is not sufficient for accurate reliability evaluation [30]. Moreover, failing to calculate/estimate the reliability starting from the device level, or simply assuming that all the gates have the same reliability, is unacceptable as it conducts to results which can be off their correct values by a few orders of magnitude [31, 216].

One of the goals of this book is to introduce a reliability evaluation method that overcomes some of the problems of existing tools, including the most significant one related to oversimplifying fault models. An MC tool can successfully be used to extract gate-level failure models, as well as for the verification of small block-level circuits. A custom-built MC tool incorporating specific failure models is described in the following section. Further developments following the reliability evaluation method development are presented in Chapter 7.

5.3 Monte Carlo Reliability Evaluation Tool

Complex redundancy schemes (multiplexing, redundancy with averaging, etc.) do not allow the extraction of a simple reliability rule, such as the majority rule applied in TMR systems, making potential analytical reliability evaluation very difficult. In a general case, every system state corresponds to an individual combination of device states that manifest themselves as degenerated output transfer function surfaces, some of which still operate correctly. In order to perform an analytical assessment of the reliability of the fault-tolerant system, it is necessary to extract a rule set which describes the combination of device states that allow correct circuit operation.

The complexity of the analytical method with complex rule sets is immense. Every constituting element of the block, such as a transistor or nanodevice, can be in a number of states dictated by the fault occurring. Calling ϵ the number of faults (variable states) and n the number of devices under consideration, the total number of system states is given as $(\epsilon + 1)^n$. For a full statistical coverage it is possible to consider a limited number of cases, given that the redundancy in the logic layer does cause a number of cases to appear as identical in their transfer function and also taken into account that faults are not totally statistically independent. Nevertheless, the actual number of states is exponentially dependent on the number of devices. Above all, the task of mapping rules into actual probabilities is tedious. Therefore, the described method could be used together with limited software support only for smaller, theoretical cases. All cases where larger Boolean networks are involved require a different approach.

In cases where the number of devices and size of blocks are limited, the use of the Monte Carlo approach provides a viable solution. A software tool using MC, SPICE, and MATLAB has been developed in order to automate the reliability analysis process for different circuit blocks, under various block sizes, redundancy factors, failure types, and a varying number of errors affecting the block.

First, a netlist is acquired from the appropriate schematic acquisition tool. Then, in each of the applied MC iterations, a faulty pattern is generated. Standard BSIM3 models [217] of the transistors that are affected by faults are replaced with the appropriate fault models according to the error model described in Section 2.3. Then a multi-variable DC sweep analysis for the acquired circuit netlist is executed, thus forming the transfer function surfaces for the considered block under failure analysis. Subsequent Monte Carlo iterations are run applying different failure patterns, performing a sweep analysis in the probability space. Subsequently, all simulation results are processed to discriminate among the faulty transfer function surfaces those which demonstrate proper circuit behavior. Finally, the related probability of failure of a gate/block/fault-tolerant circuit with respect to probability of fault of a single transistor is calculated. The described steps are shown in Fig. 5.1.

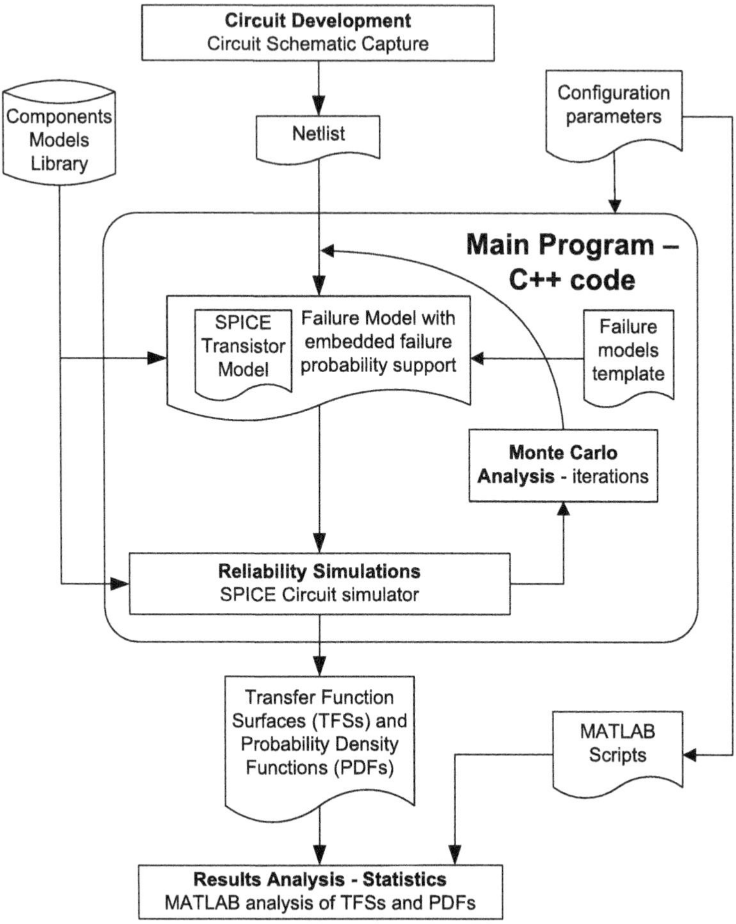

Fig. 5.1 Synthetic flow graph of the MC reliability evaluation tool

The acceptance condition for a transfer function surface to be considered as correct, despite of any errors in the circuit, can be limited to critical intervals dictated by the input noise margin of the next stage, as illustrated in Fig. 5.2b, i.e., the full search space need not be checked. The electrical meaning of the acceptance condition is depicted in Fig. 5.2a where one DC sweep for one Monte Carlo iteration is shown.

The mathematical formulation of the acceptance condition is given in (5.7). The output of the circuit is marked with V_{SGN}; V_{OH} and V_{OL} are the output noise margins, V_{IH} and V_{IL} are the input noise margins. V_{TH} is the output gate threshold value to which $\pm\Delta V_{TH}$ is attached to form a sensitivity interval:

$$\begin{pmatrix} V_{TH,H} = V_{SGN,min}|_{GND \leq V_{input} \leq V_{IL}} - \Delta V_{TH} \geq V_{TH} \\ V_{TH,H} = V_{OH}|_{GND \leq V_{input} \leq V_{IL}} - \Delta V_{TH} \geq V_{TH} \end{pmatrix} \text{ and}$$
$$\begin{pmatrix} V_{TH,L} = V_{SGN,max}|_{V_{IH} \leq V_{input} \leq V_{DD}} + \Delta V_{TH} \leq V_{TH} \\ V_{TH,L} = V_{OL}|_{V_{IH} \leq V_{input} \leq V_{DD}} + \Delta V_{TH} \leq V_{TH} \end{pmatrix} \quad (5.7)$$
$$V_{TH,range} = V_{TH,H} - V_{TH,L} \geq 0$$

Critical intervals as depicted in Fig. 5.2b, are determined by $[V_{DD}, V_{IH}]$ and $[V_{IL}, GND]$ in which the signal V_{SGN} must comply with the acceptance condition expressed in (5.7). The value of V_{SGN} outside of critical regions is not relevant.

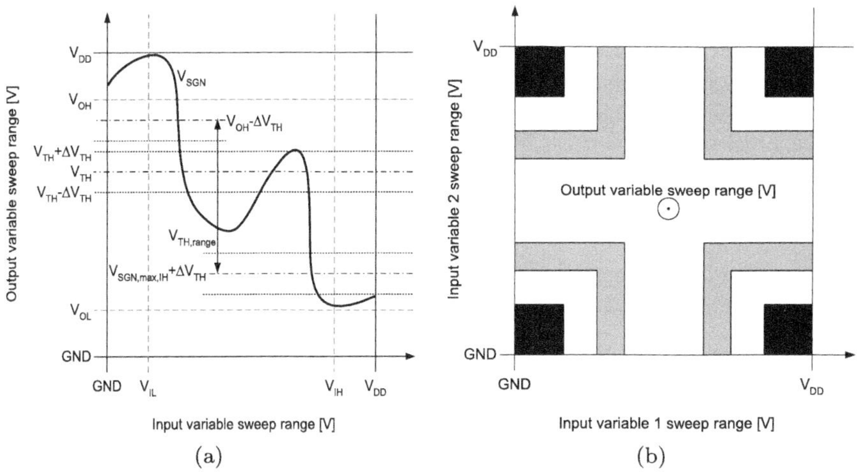

Fig. 5.2 Discrimination of correct transfer function surfaces. **(a)** Determination of V_{th} and **(b)** critical regions

The restriction of the search space to critical intervals is important because it reduces the number of input vectors to a subset that provides the satisfying fault coverage and that can easily be generated by conventional tools for automated test pattern generation (ATPG) [17]. If the fanin of the analyzed block is too high (e.g., the analyzed block has more than eight inputs) the appropriate set of test patterns providing satisfying fault coverage is automatically generated by using the commercial tool. The DC analysis is then applied only on critical intervals for each test pattern.

Fault distribution models adapted for nanometric technologies require monitoring of actual devices in mass production. The feasible models relate to the available technologies and do certainly not take into account all necessary parameters. The computational load shows an exponential dependency with the number of input variables as well as faulty states. However, in the Monte Carlo-based approach, the computational load is not exponentially dependent on any parameter; and specifically, it is not dependent on the number of fault modeling parameters. This is an additional advantage of the Monte Carlo-based approach over any analytical approach. The dependence of the single iteration time on the size of the analyzed circuit (number of states) is quadratic or smaller, and for smaller circuits linear.

The Monte Carlo approach implies state sampling. In this technique, a subset of states (sample) is randomly chosen from the set of all possible states. The states in a sample are simulated and the ratio of states where correct operation can be extracted (working states) over all states in the sample is used as an estimate of reliability in the complete state set [17]. The accuracy (or error bound) of the estimated coverage depends on the absolute number of states in the sample. This number is known as *sample size* (in our case the necessary number of Monte Carlo iterations). The error bound of the estimate can be reduced by increasing the sample size. The total number of states, N_{p}, is called the *population size*. We want to determine the population fraction R that represents the number of specific states of interest (e.g., number of working states) and randomly collect a sample of N_{s} states (sample size $= N_{\mathrm{s}}$). If r is a random variable representing probability of correct operation and x is an estimated value of R determined by Monte Carlo simulation, then the number of ways to obtain sample states N_{w} is given as

$$N_{\mathrm{w}} = \binom{RN_{\mathrm{p}}}{xN_{\mathrm{s}}}\binom{(1-R)N_{\mathrm{p}}}{(1-x)N_{\mathrm{s}}}. \tag{5.8}$$

The probability of a state sample giving a value x for the random variable r is given as

$$p(x) = \mathrm{Prob}(r = x) = \frac{\binom{RN_{\mathrm{p}}}{xN_{\mathrm{s}}}\binom{(1-R)N_{\mathrm{p}}}{(1-x)N_{\mathrm{s}}}}{\binom{N_{\mathrm{p}}}{N_{\mathrm{s}}}}. \tag{5.9}$$

This represents the *hypergeometric probability density function* of a discrete-valued random variable r. When N_{s} is large, r can be treated as a continuous variable and (5.9) is conventionally approximated by a *Gaussian probability density function* with mean, $\varepsilon(r) = R$, and variance σ^2 as expressed by

$$p(x) = \mathrm{Prob}(x \le r \le x + dx) = \frac{1}{\sigma\sqrt{2\pi}} e^{-\frac{(x-R)^2}{2\sigma^2}}. \tag{5.10}$$

Here R represents the true probability of correct operation, as the mean (or an unbiased estimate) of r. The variance of r is given according to [17] as follows:

$$\sigma^2 = \frac{R(1-R)}{N_s}\left(1 - \frac{N_s}{N_p}\right) \approx \frac{R(1-R)}{N_s}. \tag{5.11}$$

The error of estimation of r for 95% confidence interval is given as

$$|r - R| = 2\sqrt{\frac{R(1-R)}{N_s}}. \tag{5.12}$$

For example for $R = 0.5$ we only need 1,000 Monte Carlo iterations ($N_s = 1,000$) to guarantee an error smaller than 3%. This makes the Monte Carlo-based approach very suitable.

The total time for simulations to be run is expressed as follows:

$$T_{\text{sim}} = N_{\text{sp}} \cdot N_{\text{vec}} \cdot N_{\text{it}} \cdot N_{\text{prob}} \cdot T_{\text{it}} \tag{5.13}$$
$$\text{with} \quad T_{\text{it}} \propto \varepsilon^k, \quad k \leq 2.$$

Here, N_{sp} is the number of sweep points defining critical interval for each input vector, N_{vec} the number of input vectors, N_{it} the number of MC iterations, N_{prob} the number of probability iterations, T_{it} is the time of one iteration, and ε the number of possible different circuit states (circuit size). Multiprocessor systems, which optimally support parallel operation of the Monte Carlo algorithm, can be used to limit simulation time.

The described MC tool is extensively used to generate and validate results presented in Chapters 6, 7, and 8.

5.4 Summary

New reliability-enabled EDA tools are required to accurately calculate the reliability of future nanocircuits. Current generic reliability tools are not suitable because of three main reasons:

- They cannot handle the massive size of future nanocircuits.
- They customarily assume that all gates have the same reliability, which has been recently proven to introduce too wide approximations.
- They cannot be seamlessly integrate with current EDA tools for easily trading reliability vs. area/power/delay during the various design phases.

Therefore, a methodology underlying an EDA tool that is able to accurately calculate the reliability of future nanocircuits while addressing the shortcomings of the generic reliability tools needs to be developed. A first step in this methodology lies in the development of an MC tool for gate and block level probability evaluation and accurate fault modeling.

Chapter 6
Averaging Design Implementations

The fundamental element enabling reliability improvement in most of the static redundancy techniques is the decision gate, as presented in Chapter 4. One of the fundamental properties of averaging lies in the fact that it reduces the spread of output values caused by different stochastic processes, which are inherently present in hardware designs. In addition, adaptable and reconfigurable designs provide better response in situations outside of the scope or regular operation. Combining the averaging and adaptability principles into a logic circuit design can therefore significantly improve reliability.

These concepts have inspired a novel fault-tolerant technique based on averaging and adaptable thresholding of decision gates while including redundancy, which is presented in this chapter. Various implementations of this architecture are presented in Section 6.1, demonstrating the versatility of configurations where the presented technique offers improvements in reliability and yield. In Section 6.2, results of the proposed fault-tolerant technique applied to small-size blocks of standard CMOS logic are presented. The fault tolerance of circuits built from SETs is analyzed in Section 6.4.

6.1 The Averaging Technique

Methods inspired from biology can be of help in the search for fault-tolerant systems [218]. Typically, biological systems are made of numerous cells which behave as autonomous components with significant variability in their electrical characteristics, sensibility to external factors, and also comprising a number of elements that operate totally out of desirable limit performances. Moreover, living systems see their characteristics evolve in time and space, for example, under the influence of external chemical or mechanical factors. Redundancy and plasticity are fundamental properties of the central nervous system of mammals allowing learning, and adaptation, as well as providing the necessary amount of system-level robustness required to overcome deficient components or transmission lines.

The idea of analog computations has been known for many years, as it was first discussed by von Neumann [13]. Recent works [26, 138, 139, 141] suggest

M. Stanisavljević et al., *Reliability of Nanoscale Circuits and Systems*,
DOI 10.1007/978-1-4419-6217-1_6, © Springer Science+Business Media, LLC 2011

the use of averaging and thresholding as an interesting research topic. The perceptron (Fig. 6.1) as an artificial neural network (ANN) in a general sense is a classical implementation of an averaging and thresholding gate [136, 140, 141, 219]. A fault-tolerant technique inspired by the ANN realization of weighted averaging and thresholding is the main subject of this chapter. Therefore, more insight into ANN realization is given further in the text.

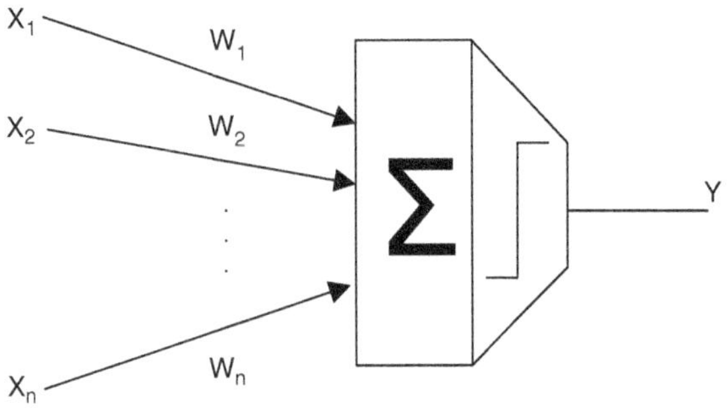

Fig. 6.1 Perceptron (threshold element). The input signals X_1, X_2, ... have weights W_1, W_2, ... If the weighted sum of inputs exceeds the threshold T, the binary output Y changes

6.1.1 Feed-Forward ANN Boolean Function Synthesis Block

A possible direction that is proposed by several researchers [15, 132] for the realization of complex functions using nano-scale devices involves the design of artificial neural network (ANN) architectures [220]. The main functionality of neural building blocks is based on the computation of synaptic weighted-sum terms on the output node. While offering some significant advantages in terms of system-level robustness, the classical ANN approach also has several hurdles associated with its implementation. The elementary synaptic functions can be realized by using nanodevices (specifically SETs), but the hardware realization of learning algorithms remains a problematic issue. Also, the interconnect density in general-purpose ANN architectures represents a serious limitation in terms of hardware complexity, which is exacerbated by the introduction of general-purpose learning/adaptation hardware. Finally, the classical ANN paradigms based on learning algorithms and self-adapting hardware are in extreme contrast to the conventional, well-established system design methodologies currently being used for CMOS VLSI systems. In the following, we explore a design approach that borrows some of the fundamental aspects of ANN architectures such as the replication of identical hardware blocks, the low-level regularity, multiple functional layers, and regular, short-distance interconnections.

The reliability of ANNs has been tackled by several research groups [221, 222]. The algorithmic aspects, as well as the hardware implementation issues have been addressed, mainly with the target of guaranteeing the correct ANN transfer function under failure, while considering a minimal number of neurons [223–225]. Also the aspects of learning using error-prone analog hardware and/or limited precision digital circuits have been addressed. In our approach, feed-forward ANNs (FFANNs) are used as a tool to implement Boolean functions, where each function implementation is perceived by the designer as a black box. An appropriate training technique, as well as the appropriate neuron redundancy is used to provide enhanced fault tolerance.

The key issue is to design each functional block with the capacity of generating, transmitting, and receiving analog levels and to define "regions of acceptability" that comply with this concept, instead of using the classical noise margins defined in digital design. The output of any function block may be varying depending on its state, on possible cross talk perturbation, and on the actual operation of the basic transistors, which may exhibit unreliable behavior, such as soft and/or unpredictable cut-off or turn-on levels. Rather than handling these issues with noise margin intervals, it is more appropriate to design the input stages to be compliant with "regions of acceptability." Hence, one possible solution would consist of designing each functional module with two complementary analog outputs, where each output is considered as a degree of confidence to be given to logic high and logic low levels, respectively. Figure 6.2 depicts a possible realization, covering a three-layer FFANN architecture with analog, complemented inputs and outputs which is designed to perform a simple Boolean operation (NAND, NOR, etc.).

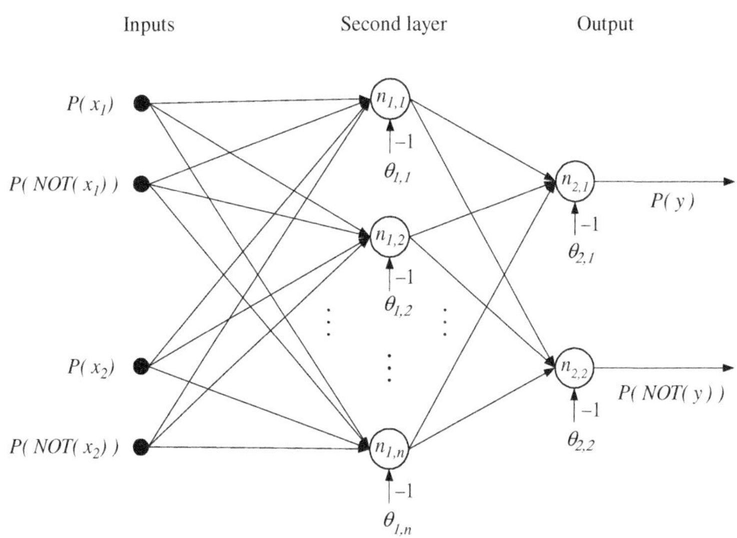

Fig. 6.2 Three-layer FFANN with analog, complemented inputs and outputs, designed to perform a simple Boolean operation

6.1.2 Four-Layer Reliable Architecture (4LRA)

A design approach that borrows some fundamental aspects of ANN architectures such as the replication of identical hardware blocks, the low-level regularity, the existence of multiple functional layers, and the implementation of regular short-distance interconnections is explored. At the same time, the proposed approach avoids more problematic aspects such as ANN learning paradigms and adjustable weights, concentrating on the robust realization of Boolean functions that can absorb a certain degree of random device failures within a functional block.

The proposed four-layer fault-tolerant hardware architecture (4LRA) (Fig. 6.3), which is named in agreement with the layer-naming conventions in ANNs, is based on a four-layer feed-forward topology and uses averaging and thresholding as the core voter mechanisms [135, 137, 226, 227]. The architecture can have a fixed or adaptable threshold and is applicable at the gate or extended gate level. It can be applied hierarchically in a bottom-up way and combined with other high-level fault absorption techniques.

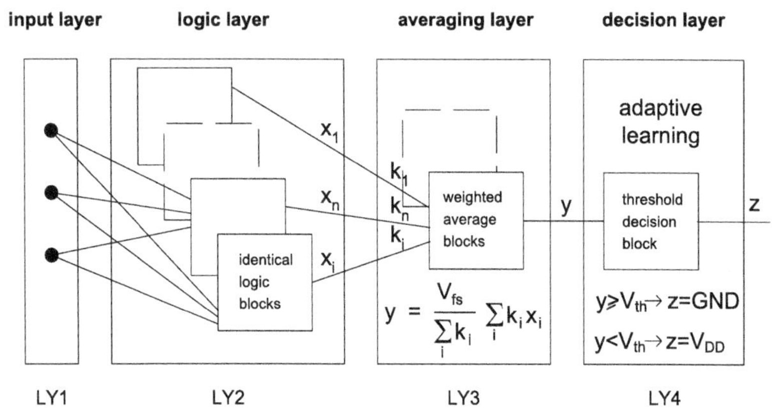

Fig. 6.3 The fault-tolerant architecture based on multiple layers

The proposed fault-tolerant architecture consists of four layers in which data are strictly processed in a feed-forward manner (Fig. 6.3). The first layer is denoted as the input layer, accepting conventional Boolean (binary) signal levels. The core operation is performed in the second layer, which consists of a number of identical, redundant units each implementing the desired logic function. The fault immunity increases with the number of redundant units, yet the operation is quite different from the classical majority-based redundancy. The third layer receives the outputs of the redundant logic units in the second layer, creating a weighted average with rescaling to match the full range of signals (e.g., in the voltage domain). Note that the output of the third layer becomes a multiple-valued logic level. Finally, the fourth layer is the decision layer where a binary output value is extracted using a simple threshold function.

The block named LY3 in Fig. 6.3 shows the typical weighted-averaging and rescaling function that is performed by one of the third-layer blocks. The benefits of this circuit architecture can be further increased by using multiple averaging units and/or different input weighting schemes. The acceptance condition for a transfer function surface of the circuit that is operating correctly (an example for 2-input NAND gate depicted in Fig. 6.5) is dictated by the possibility to place a threshold value V_{th} in a way that permits a correct separation of the logic-1 and logic-0 outputs. Considering only the pull-down network that realizes a multi-variable Boolean function, the condition for correct restoration of the output function under multiple device failures can be expressed as

$$V_{\text{NM}} \leq \min_{\substack{\forall V_{j \in \mathbb{L}} \\ \forall V_{\text{out}\,i \in \mathbb{P}}}} \left\{ \frac{1}{2} \left(V_{\text{H min}} - \frac{V_{\text{fs}}}{\sum_{i=1}^{R}} k_i V_{\text{out}\,i} (V_j \mid j = 1, \ldots, m) \right) \right\}, \quad (6.1)$$

where V_{NM} is the allowable output noise margin, $V_{\text{H min}}$ is the lowest level for logic-1 output, V_{fs} is the nominal full-scale range for the output voltage, k_i is the weight coefficient, $V_{\text{out}\,i}$ is the actual output voltage of each unit in the second layer under the assumption of device failures, V_j is the input voltage vector, and R is the number of redundant units in the second layer. Note that $\mathbb{L}$ represents the set of all m input combinations that are expected to produce a logic-0 output in the undamaged gate and $\mathbb{P}$ represents the set of all outputs for each unit in the second layer, with or without device failures. It is assumed that the threshold point in the fourth layer is at $V_{\text{DD}}/2$, where V_{DD} is the supply voltage. A dual expression, similar to the one given above, can also be derived to define the conditions for the pull-up network.

Correct binary output of the proposed failure-absorbing architecture is obtained by thresholding the output of the third layer. The fourth layer consists of a hard-limiting thresholding module, which makes a decision based on the transfer function surface. The actual value of the threshold is a key parameter for proper operation of the system. The correct value of the threshold can be extracted from analyzing the transfer function surface. This process is straightforward for a human operator working on a single gate. However, real implementations of averaging and thresholding layer are necessary for the fabrication of fault-tolerant designs.

The fault-tolerant circuit design strategy described above can be used to construct simple standard-cell-like building blocks with a high level of functional immunity against possible permanent and transient device failures.

Due to the importance of the proposed technique based on averaging and adjustable thresholding that forms a central focus in further chapters, a detailed evaluation of the performance of this novel fault-tolerant technique is presented in Sections 6.2 and 8.1.

6.1.3 Hardware Realizations of Averaging and Thresholding

The averaging operation can be realized in different ways. The most typical realizations are explored in the following.

The most common way would consist of converting the averager input voltages into the current domain, applying simple summation of currents and performing a current-to-voltage conversion in the final stage. The typical realization is described in Section 6.2 and an improved differential realization is presented in Section 6.3. The serious drawback of this realization relates to its static current requirement, which leads to static power consumption. A large number of averagers on the chip can significantly increase the power consumption which is not acceptable in large modern circuits.

A capacitive threshold logic (CTL) realization described in [136] does not suffer from power consumption problems and is reported to be fast (200 MHz in a 0.35 μm technology). However, CTL operates as a dynamic logic which is not standard today and suffers from synchronization problems causing a high sensitivity to noise and process variations.

The most recently proposed circuits are a CMOS floating gate realization [228] and configurations that exploit the transistor as a four-terminal device [141, 229]. These configurations are fast, have low power consumption, are suitable for low-power applications, and offer a viable solution for a possible averaging realization. Moreover, the threshold value of the circuits can easily be controlled by an external analog control voltage. However, the linearity of averagers realized in these configurations needs to be improved to permit efficient future applications in fault tolerance.

Automatization of the threshold setting process is required in order to propose a method that can effectively be applied to realistic circuits. Two possible approaches are proposed to solve this critical issue:

- an approach based on the theory of artificial neural network training and
- an approach based on reconfiguration.

6.1.3.1 Neural Network Thresholding Realization

One fundamental prerequisite is expressed as the fact that a complex Boolean operator can be broken down into clusters consisting of simple functions where the dataflow direction is in a strict feed-forward path and where the input–output ranges are known. Under these conditions, it is possible to consider the simpler function derived in the preceding step as an artificial neural network which must be trained in order to perform a predetermined function. Here, every neuron is represented as a Boolean gate having the threshold as the only adaptable parameter and a hard-limiting function as the activation function. Learning can be applied to these units using an adapted version of the backpropagation learning rule [230] or weight-perturbation learning [231] and the threshold level properly adapted to absorb any possible errors in the first three layers, under the condition that the fourth layer be spared from any fault. A chip-in-the-loop arrangement of the training system with respect to already trained system seems to be appropriate for this approach, which requires systematic testing under the constraints of a predefined set of test patterns consisting of input-expected output pairs of data.

The training process described in the preceding paragraph requires devoting a long time to system configuration prior to any computation. Moreover, it is capable to correct permanent errors only. Dynamic adaptation of the threshold aimed at absorbing transient errors requires a control and correction process to be applied in real time to every gate. Weight-perturbation algorithms can be applied to update the weights on the fly. However, this requires to integrate the training sets and some control on-chip, which could be costly in terms of power, area, delay. Hence, the granularity of the hardware to be trained results from a trade-off with the extra hardware that must be incorporated, resulting in a solution which should consist of a combination between the chip-in-the-loop and the actual on-chip learning training methods.

6.1.3.2 Reconfiguration-Based Thresholding Realization

Another method to perform optimal threshold setting consists of a reconfiguration-based approach. A concept of the possible solution is presented in the following.

A system that consists of reliable blocks where the four-layer reliable architecture has been applied is depicted in Fig. 6.4. All outputs of reliable blocks (blue wires) are connected to the so-called reliability control unit. Moreover, each thresholder in 4LRA has controlling inputs (red wires). Controlling inputs are used for setting an appropriate output threshold. A suitable averager/thresholder circuit can be a previously described floating gate [228] or four-terminal transistor realization [141, 229].

The reliability control unit can be realized as a series of scan chains which can be turned off during normal chip operation, and thus are not consuming power. Reconfiguration can automatically be performed in the chip-testing phase. The reconfiguration includes

- the identification of the defective blocks/outputs by means of classical chip testing [17];
- the configuration of the thresholds of defective blocks/outputs by exploring the whole available space of configurable values of the threshold; as far as the number of defective blocks/outputs is not large, this process can be effectively performed within the common testing time [17].

An advantage of this approach would be observed as a much faster configuration than in a neural network learning approach. However, the drawback is the same: the dynamic adaptation of the threshold aiming at absorbing transient errors requires a control and correction process to be applied in real time to the reliable block.

The reliability control unit should be fault free. This hypothesis is acceptable considering that the reliability control unit is very small compared to the operative part of the circuit, and thus can be hardened.

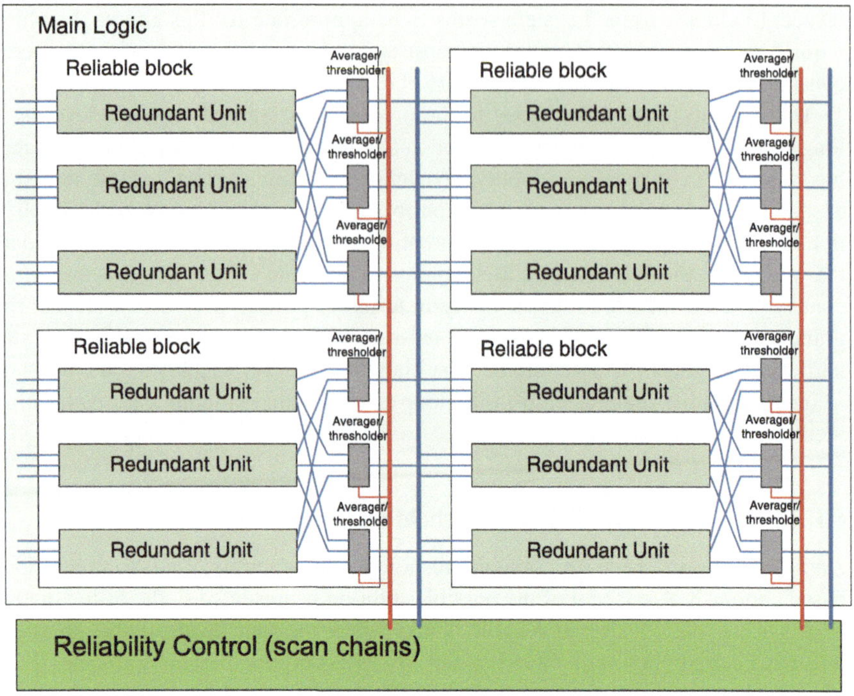

Fig. 6.4 Conceptual schematic of the reconfiguration-based thresholding

6.1.4 Examples of Four-Layer Reliable Architecture Transfer Function Surfaces

To demonstrate the effectiveness of the proposed fault-tolerant architecture and to examine its immunity against multiple device failures, some simple examples are presented. The 2-input NOR function is selected as a test case; other Boolean primitives or more complex functions can also be implemented using the same strategy.

The first example consists of a 4LRA architecture with two identical logic blocks in the second layer and one averaging block in the third layer. It is assumed that the NOR function blocks in the second layer are realized as standard CMOS circuits. Each NOR block in the second layer receives two binary inputs and produces one binary output. The outputs of the second layer are further processed in the averaging block to produce the multiple-valued output. As long as all devices operate correctly, three of the four possible input combinations produce a logic-0 output in both of the second-layer logic blocks and only one input combination "00" produces a logic-1 output. Figure 6.5 shows the output level of the averaging block for correct operation. The transfer function surface generated at the output of the averaging block clearly reproduces the expected 2-input NOR function, with the fourth-layer decision threshold set as shown in solid. The fundamental characteristics of the

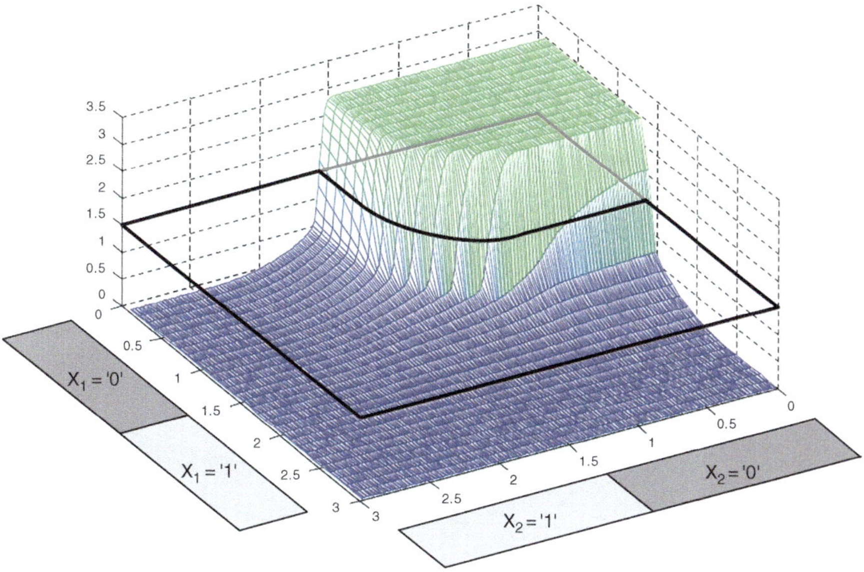

Fig. 6.5 Output transfer function generated by the averaging layer (output of the third layer) of the 2-input NOR circuit with two redundant units, showing correct operation (no device failures). The fourth-layer decision threshold is also indicated as the *black wireframe*

proposed architecture are largely independent of the specific implementation of the logic function blocks and the averaging block [232].

Only two types of simple faults are considered for the devices in all layers, namely stuck-on and stuck-off. Obviously, more refined fault models, as the ones presented in Section 2.3, can be used to reflect additional, more subtle kinds of failures. Possible failure modes due to interconnects between the gates are not taken into consideration in this analysis.

Considering random device failures, the proposed architecture can successfully absorb all single faults occurring anywhere in the second layer, as long as two or more identical logic units are implemented in the second layer. This is a property that can only be achieved using three or more redundant units in the conventional approach based on majority voting. Furthermore, the new circuit architecture is capable of producing correct output behavior even when some devices in the third layer (averaging block) are faulty. This is in contrast to the limited fault immunity of conventional redundant systems where even a single fault in the majority decision block cannot be tolerated. The significant benefits of the proposed design approach become evident especially when considering multiple device failures. Figure 6.6 shows the output transfer function surface of the circuit described above (two identical NOR blocks in the second layer, one averaging block in the third layer) where a total of four devices are assumed to be faulty. The correct output behavior can be extracted by setting the decision threshold level as shown. A fixed decision threshold

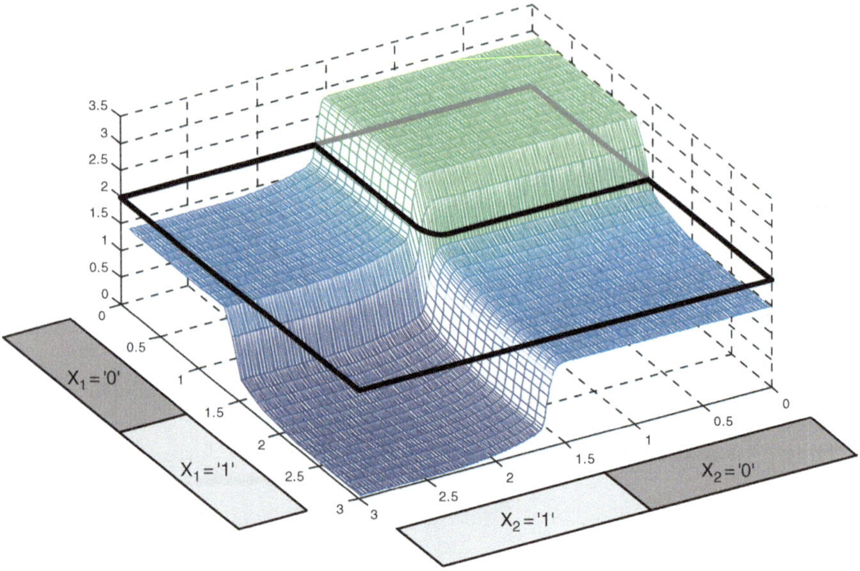

Fig. 6.6 Output transfer function generated by the averaging layer (output of the third layer) of the 2-input NOR circuit with two redundant units, assuming a total of four device failures in both the second-layer logic blocks. The correct output can still be obtained with the proper threshold decision in the fourth layer (indicated as the *black wireframe*)

level appears to be sufficient in most cases, while dynamically adjustable decision threshold levels may further increase the flexibility of the proposed approach.

Figure 6.7 shows the third-layer output transfer function surfaces of the NOR circuit with two redundant logic units, under the assumption of several different (single or multiple) device failures. The multiple-valued output level of the averaging block is capable of preserving the essential function in most of the cases, and the correct binary output can be extracted by applying a simple decision threshold, however, at the expense of reduced noise margins. As mentioned earlier, the fault tolerance of the proposed circuit architecture increases with the number of identical, redundant logic blocks used in the second layer. Figure 6.8 shows the output transfer surfaces of the NOR circuit with three redundant logic units in the second layer, again assuming single or multiple device failures.

The graceful degradation of the transfer function surface due to various injected faults suggests that the proposed circuit architecture is capable of absorbing a large variety of faults and still producing the correct binary output. In comparison to the classical triple redundancy with majority voting, note that the proposed approach is capable of withstanding single device failures using two redundant units only (redundancy factor of $R = 2$). Also, the correct output function can still be obtained in a large number of cases with multiple device failures. This would not be the case for triple redundancy with majority voting, where the probability of correct operation drops very sharply with the number of device failures [111]. The capability to

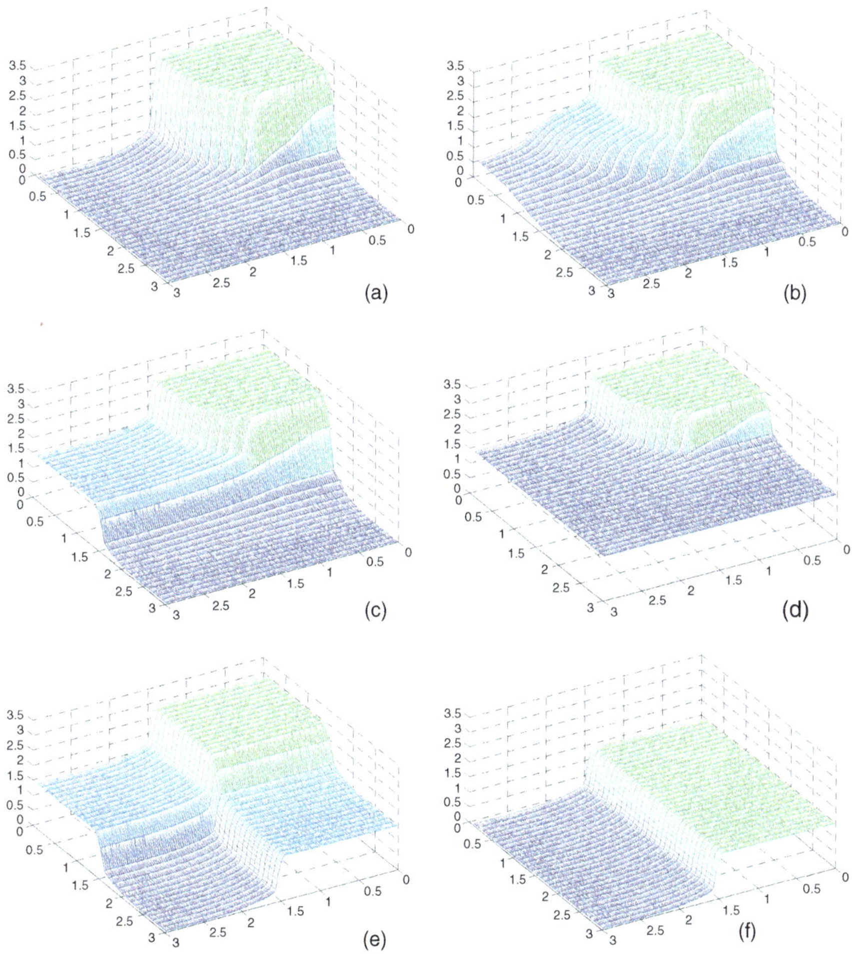

Fig. 6.7 Output transfer function of the 2-input NOR circuit, with only two redundant units: **(a)** no device failure, **(b)** one failure in the second layer, **(c)** two failures in the second layer, **(d)** one failure in the third layer, **(e)** four failures in the second layer, and **(f)** four failures in the second layer. Except in case **(f)**, the correct output function can still be obtained by applying the proper threshold decision in the fourth layer

withstand multiple device failures with a small number of redundant units would be especially valuable in cases with very high defect density, where the classical triple redundancy scheme has very limited utility. The fault-tolerant design approach described here is not limited to the mappings of two input variables onto one output variable and can be extended to the cases where more input and/or output variables are involved. These cases require adaptive systems, where the output thresholds should be multiple and should be adjusted in real time [232].

The level of immunity against device failures that is provided by using the proposed circuit design approach can be quantitatively demonstrated by considering

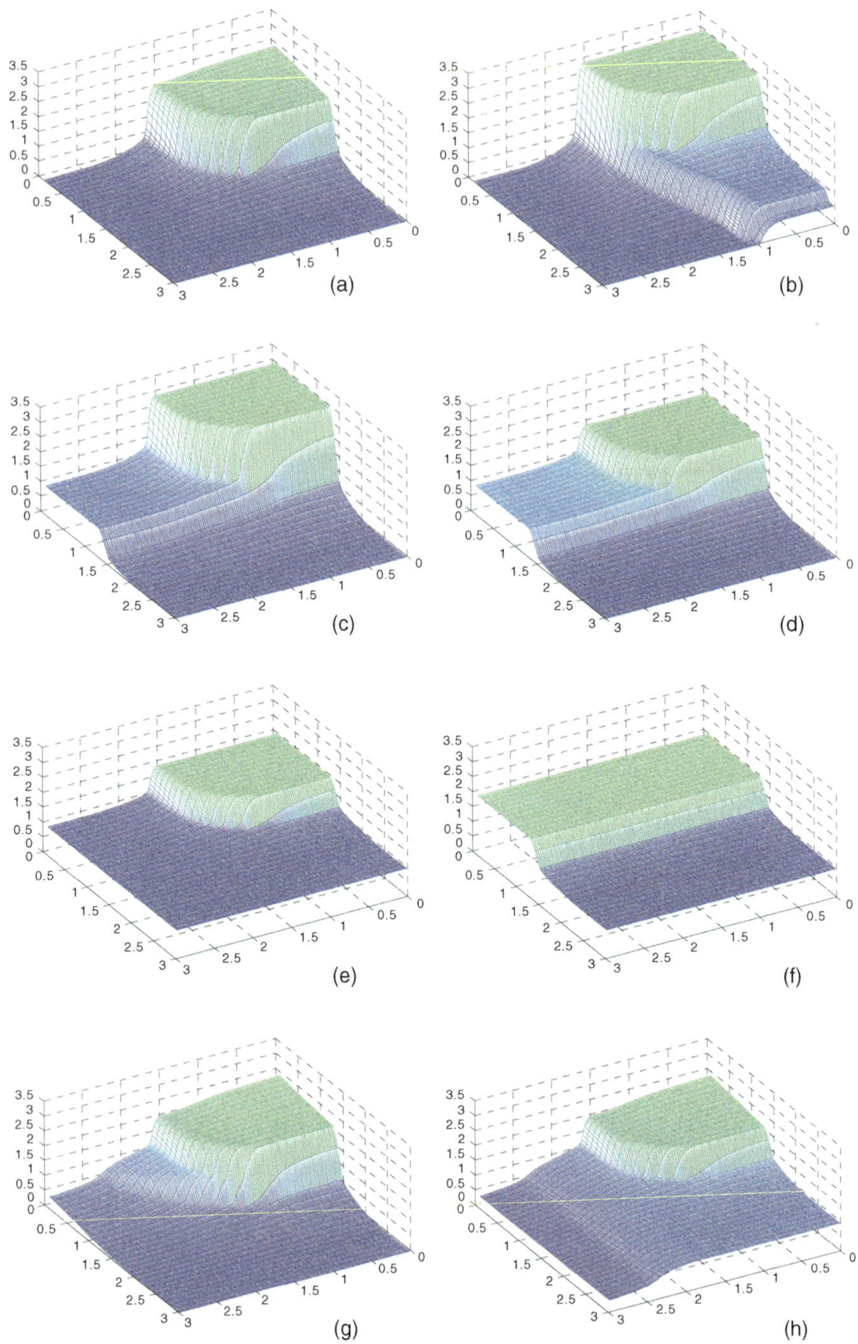

Fig. 6.8 Output transfer function of the 2-input NOR circuit, with three redundant units: (**a**) one device failure in the second layer, (**b**) one failure in the second layer, (**c**) two failures in the second layer, (**d**) four failures in the second layer, (**e**) two failures in third layer, (**f**) four failures in the second layer and one failure in the third layer, (**g**) two failures in the second layer, and (**h**) four failures in the second layer. Except in case (**f**), the correct output function can still be obtained by applying the proper threshold decision in the fourth layer

the overall probability of correct operation, as a function of the device failure probability in that particular function block. For this analysis, we first examine the realization of the 2-input NOR function described in the previous section, based on two identical redundant logic blocks in the second layer (i.e., $R = 2$) and one averaging block in the third layer. It is assumed that one or more devices suffer a stuck-on or stuck-off fault and that the failure(s) can occur in a completely random manner in any of the redundant logic blocks. All single device failures and all multiple device failure combinations have been exhaustively tested, in order to provide a realistic assessment of the circuit reliability. Possible device failures affecting the averaging block or the threshold decision layer have not been considered in the following analysis.

Figure 6.9 shows the probability of correct operation of the 2-input NOR circuit with two redundant units, as a function of the device failure probability. Correct circuit operation can be obtained with a random device failure probability as high as 0.1. This would already correspond to a high defect density. The probability of correct operation gradually drops for higher device failure rates; yet the tendency of this deterioration clearly shows a graceful degradation rather than a sharp decline. It is worth noting that using only two redundant units would not provide any immunity against single faults in the conventional majority voting approach, where a minimum of three redundant logic units and a fault-free majority decision block would be needed to achieve a comparable level of immunity.

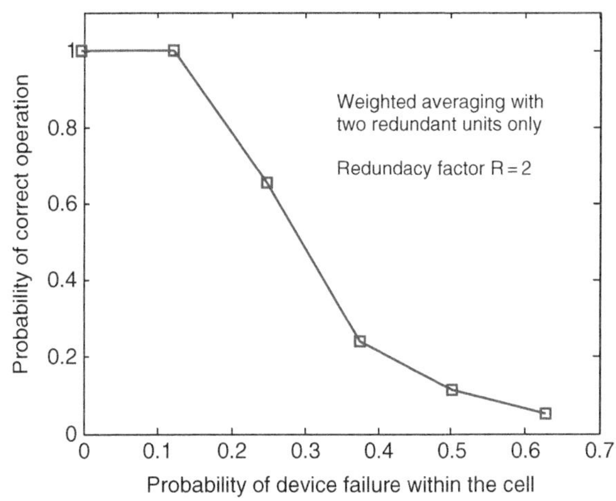

Fig. 6.9 Probability of correct operation of the 2-input NOR circuit with two redundant units in the second layer, as a function of the device failure probability. The curve was obtained with exhaustive application of all possible device failure scenarios

The level of fault immunity can be further increased using three identical logic blocks ($R = 3$) in the second layer, as shown in Fig. 6.10. In contrast to the previous case, triple redundancy with majority voting can ensure correct operation for a device failure rate up to 0.08, and the probability of correct operation sharply drops for higher device failure rates. Hence, the proposed circuit design approach based on weighted averaging of redundant units is shown to provide a significant level of immunity against permanent or transient random device failures. Also, note that the probability of correct operation can be further improved by increasing the number of redundant logic blocks in the second layer.

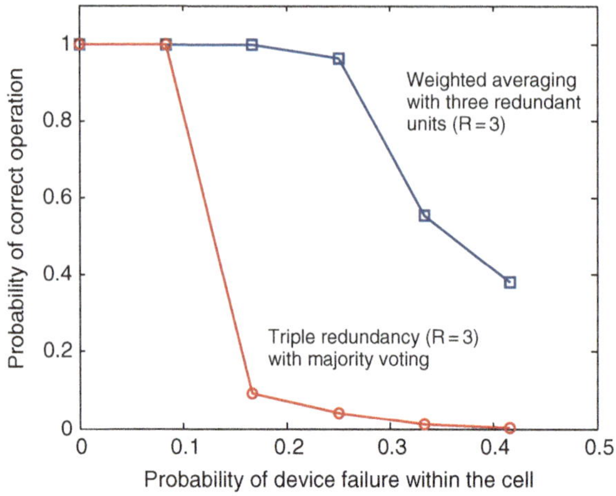

Fig. 6.10 Probability of correct operation of the 2-input NOR circuit with three redundant units in the second layer, as a function of the device failure probability. The curve was obtained with exhaustive application of all possible device failure scenarios

6.2 Assessment of the Reliability of Gates and Small Blocks

In the process of detailed assessment of the performance of fault-tolerant architectures, different types of circuits (basic gates, Boolean, as well as more complex circuits as full adders) and different fault-tolerant techniques (RMR with majority voting and different averaging schemes) with various redundancy factors have been extensively analyzed, using the software tool presented in Section 5.3. A detailed transistor fault model (presented in Section 2.3) has been used. The following fault-tolerant techniques are considered:

- R-fold modular redundancy with a majority voter, where R redundant units feed a decision gate represented by a majority voter implementing a $(R + 1)/2$ out of R majority function. This technique is named RMR in the following.
- Four-layer reliable architecture where the fourth layer is implemented as a fixed threshold layer, with the threshold point at half of the supply voltage ($V_{th} = V_{DD}/2$). This technique is named AVG in the following.
- Four-layer reliable architecture where the fourth layer is implemented as an adjustable threshold layer which is set in advance to an optimum threshold point

according to probability of occurrence of logic-0 and logic-1. This technique is named AVG-opt in the following.
- Full four-layer reliability architecture where the fourth layer is implemented as an adaptable threshold that is set to an optimum threshold for each fault pattern that is detected. This technique is named 4LRA in the following.

Each of these fault-tolerant techniques is analyzed in two configurations, namely, considering that (i) decision gates (third and fourth layer) are fault free and (ii) the input transistors of decision gates are affected by the same defect density as the rest of the circuit.

The averaging function processed in the third layer of 4LRA is a key component of the presented fault-tolerant architecture. The circuit performing the averaging function has been designed for different numbers of inputs and has been developed based on the current mode of operation. A realization of the circuit with four inputs is depicted in Fig. 6.11. The input stage is composed of R PMOS transistors (for R-input averaging function) operating in either cut-off or linear (non-saturated) regimes. Each transistor gate is driven by the analog voltage level produced in the second layer. The NMOS current mirror replicates the input stage current to the next stage. The averaging function is realized by the voltage drop across the resistor. The proposed circuit is not expected to operate in continuous analog domain. The main design constraint dictates that the output level be inside one of R acceptance intervals, resulting from the switching on or off of the PMOS transistors. The magnitude of the acceptance intervals is imposed by the desired noise margin as explained in Section 5.3. The biasing should guarantee that input transistors are in linear (non-saturated) regime so that linearity is maximized. Still, when input signals are within the $[V_{DD} - V_{PT},\ V_{DD}]$ interval (V_{PT} is the threshold voltage for PMOS), input transistors are off and the averager output is insensitive to its inputs. Also note that the averaging unit described here can be built with regular CMOS devices rather than nanometer-scale devices, since each averaging circuit actually serves a large number of function blocks in the second layer.

The adaptable thresholding in the fourth layer of 4LRA is ideally performed by the tool (Section 5.3, Fig. 5.3) for all circuits that are analyzed in this chapter, rather than being realized by the actual circuit.

6.2.1 Comparative Analysis of Obtained Results

A comparative analysis of all considered fault-tolerant techniques with a *fault-free* decision gate and considering different redundancy factors is presented in Fig. 6.12. For $R = 3$, the use of the 4LRA enables correct circuit operation with a probability of 90% over a plateau extending to 37% of the probability of transistor fault, and gracefully degrades, where RMR shows a clearly weaker resistance to transistor fault. Averaging with fixed threshold (AVG) also shows better performance than RMR. Averaging with adaptable threshold (AVG-opt) shows a performance that is

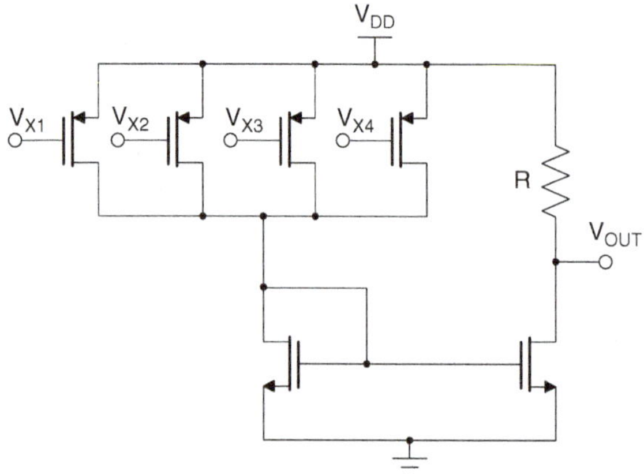

Fig. 6.11 Single-ended realization of the averager [233]

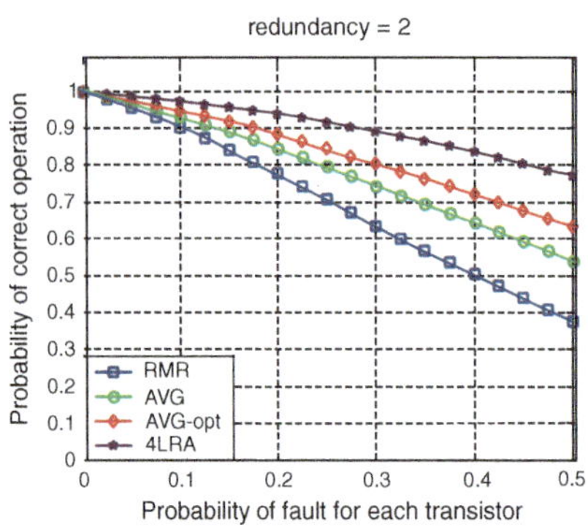

(a) redundancy factor $R = 2$.

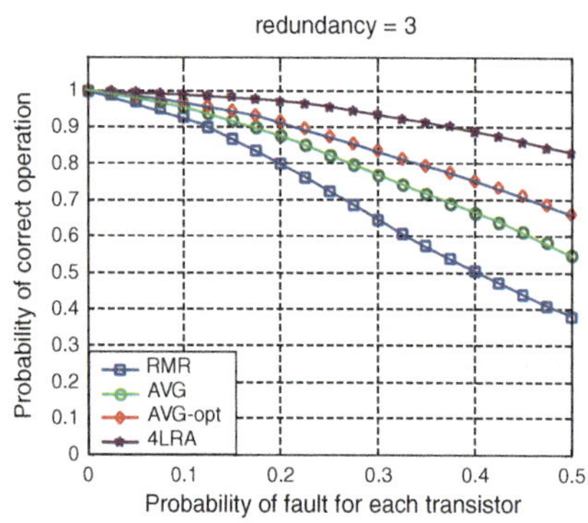

(b) redundancy factor $R = 3$.

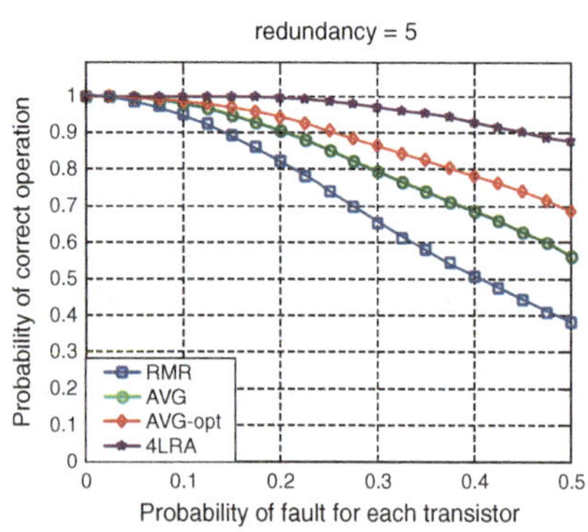

(c) redundancy factor $R = 5$.

Fig. 6.12 Comparative analysis of the 2-input NAND gate in RMR, AVG, AVG-opt, and 4LRA fault-tolerant configuration with a *fault-free* decision gate and for redundancy of $R = 2$, 3, and 5

comparable to 4LRA and where correct circuit operation with a probability of 90% is possible over a plateau reaching out to 18% of the probability of transistor fault.

For other redundancy factors, the relative differences between different fault-tolerant techniques remain as presented, but the absolute reliability figures change. For $R = 5$ the use of the 4LRA enables correct circuit operation with a probability of 90% over a plateau extending to almost 45% of the probability of transistor fault, and for $R = 2$ this drops to below 30% for 4LRA and less than 10% for RMR.

When faults are induced into the decision gate (averager and majority voter input transistors are affected) the performance is significantly degraded, as depicted in Fig. 6.13. Still, for $R = 5$, the use of the 4LRA enables correct circuit operation with a probability of 90% over a plateau reaching out to 30% of the probability of transistor fault. For lower redundancy factors, the degradation in performance is stronger.

When larger blocks are used as redundant units, such as a gate realizing the complex 4-input function $f(x1, x2, x3, x4) = x1x4' + (x2x3)' + x1(x2x3)' +$

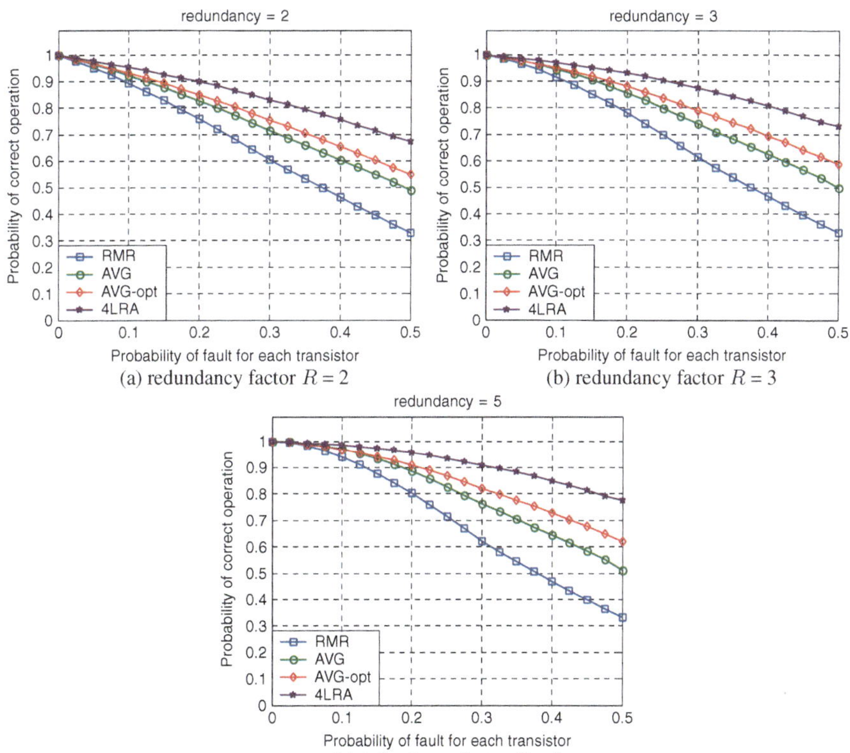

(a) redundancy factor $R = 2$

(b) redundancy factor $R = 3$

(c) redundancy factor $R = 5$

Fig. 6.13 Comparative analysis of the 2-input NAND gate in RMR, AVG, AVG-opt, and 4LRA fault-tolerant configuration with *faulty* decision gate and for redundancy of $R = 2, 3$, and 5

$x1'x2x3x4$ and a full adder (FA) cell realized as mirror adder, the probability of correct operation is reduced with respect to the size and complexity of a block (order of size or complexity: 2-input NAND, C_{out} output of a FA, S output of a FA, 4-input function). This is depicted in Figs. 6.14 and 6.15. Nevertheless, again an advantage of 4LRA in comparison to RMR is evident. With the increase of redundancy, a clear improvement in performance can be observed in Fig. 6.14.

The impact of using a *faulty* decision gate compared to a *fault-free* decision gate is reduced when the block size or complexity is increased, which can be noted by comparing Figs. 6.12, 6.13, and 6.15.

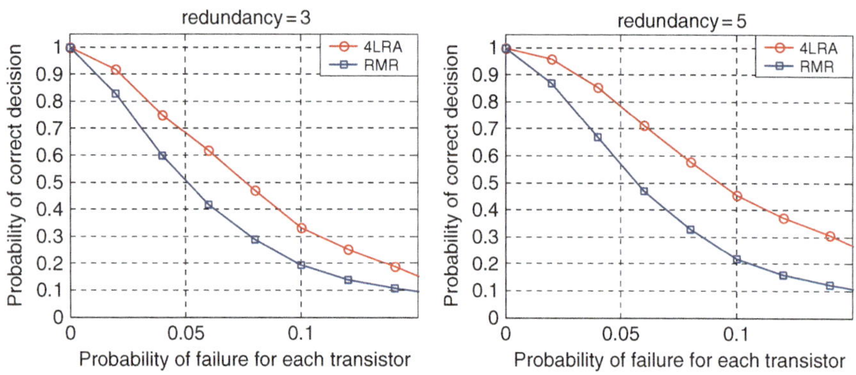

(a) redundancy factor $R = 3$ (b) redundancy factor $R = 5$

Fig. 6.14 Comparative analysis of the 4-input gate realizing the function $f(x1, x2, x3, x4) = x1x4' + (x2x3)' + x1(x2x3)' + x1'x2x3x4$ in RMR and 4LRA fault-tolerant configuration with *faulty* decision gate and for redundancy $R = 3$ and $R = 5$

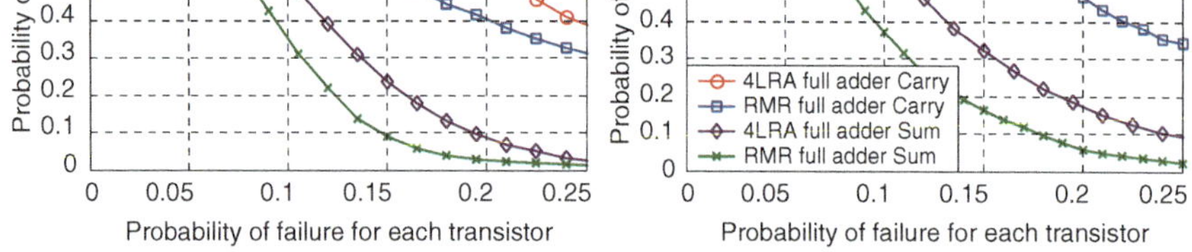

(a) fault-free model of averager circuit (b) faulty model of averager circuit

Fig. 6.15 Comparative analysis of the full adder cell in RMR and 4LRA fault-tolerant configuration for redundancy $R = 3$ in case of *fault-free* and *faulty* decision gate

6.3 Differential Signaling for Reliability Improvement

6.3.1 Fault-Tolerant Properties of Differential Signaling

In differential circuits, information is processed and transmitted in a redundant and complementary way, intrinsically offering an increased resistance to failures. In case of failure, the correct output signal can still be recovered, if (i) the complement signal is available and (ii) the circuitry which can decode this state is available. A strong motivation pertains to exploring the reliability of fault-tolerant architectures that exploit differential signaling.

The logic decision in the thresholding element (gate or voter) is based on the possibility to define a decision interval centered around a threshold value $[V_{th} - \Delta V_{th}, V_{th} + \Delta V_{th}]$ separating the complementary output line voltage values representing logic-1 and logic-0. The exact values of V_{th} and ΔV_{th} are dictated by the circuit construction of the next layer input stage. Nevertheless, some conclusions with significant practical impact can be drawn out of the theoretical rationale.

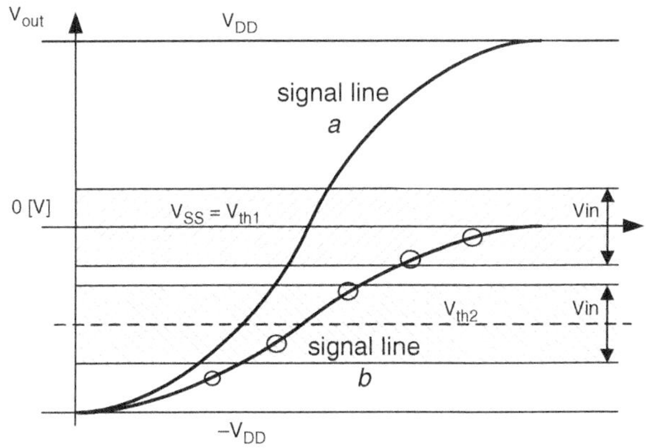

Fig. 6.16 Effect of stuck-at errors on the transfer function and corresponding adaptive value of V_{th}: (1) failure free (*plain line*, V_{th1}), (2) stuck-at-zero (*circled line*, V_{th2})

The impact of a stuck-at fault is perceived on the transfer function surface as a compression of the analog output range, as depicted in Fig. 6.16. Consequently, a variable decision threshold V_{th} is mandatory in order to handle all possible combinations of failure distribution. Let the output of layer three of 4LRA be complementary lines a and b. Calling the values on the lines a and b corresponding to logic-1 and logic-0 a_1, b_1 and a_0, b_0, respectively, the value of V_{th} which is appropriate to handle errors is the arithmetical average of differential signals, taking actual values into account, i.e., any stuck signal is assigned its actual stuck voltage value and is expressed as

$$V_{th} = \frac{(a_1 - b_1) + (a_0 - b_0)}{2}$$

A thresholding circuit complying with this property is applied in the four-layer architecture and used in the analysis in Section 6.3.2. The averager (layer three of 4LRA) is designed to be compatible with differential signaling, having differential inputs and outputs (Fig. 6.17). The differential output of the circuit is already in the voltage domain, hence there is no need for output current-to-voltage conversion. The linearity of this circuit is improved compared to the single-ended realization (Fig. 6.11) due to absence of intervals when all transistors are off, as in the single-ended realization. Using the differential averager, only the gain is reduced when input signals are within the $[V_{DD} - V_{THP}, V_{DD}]$ interval, where V_{THP} is the PMOS transistor threshold voltage. However, the condition of biasing remains the same as in the single-ended version. The biasing should guarantee that input transistors are in linear (non-saturated) regime. Compared to single-ended realization, the differential realization offers better linearity and faster operation, but occupies a larger area.

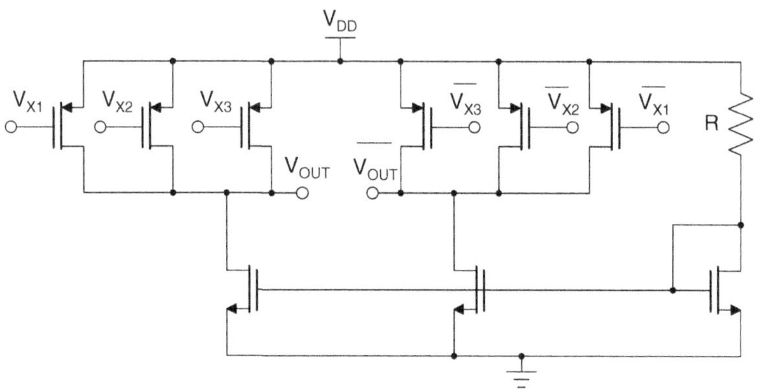

Fig. 6.17 Differential-ended realization of the averager [233]

6.3.2 Comparative Analysis of Obtained Results

Differential cascode voltage switch (DCVS) logic [234] is a differential circuit technique which has potential advantages over conventional static CMOS logic in terms of circuit delay, layout density, power dissipation, and logic flexibility. In this section, we demonstrate key advantages in terms of reliability of DCVS logic in comparison to standard CMOS logic, both used in a full four-layer reliable architecture configuration (4LRA). Different types of circuits (basic gates, Boolean, as well as more complex circuits as full adders) and different circuit topologies (standard CMOS logic, static DCVS) with various redundancy factors have been extensively analyzed, using the software tool presented in Section 5.3. The reliability figure is defined as the probability of correct operation with respect to the probability of failure of each transistor. Some basic DCVS gates are depicted in Fig. 6.18.

The comparative analysis of CMOS and DCVS logic realizations of a 2-input NOR gate is depicted in Fig. 6.19. In both topologies, the blocks performing

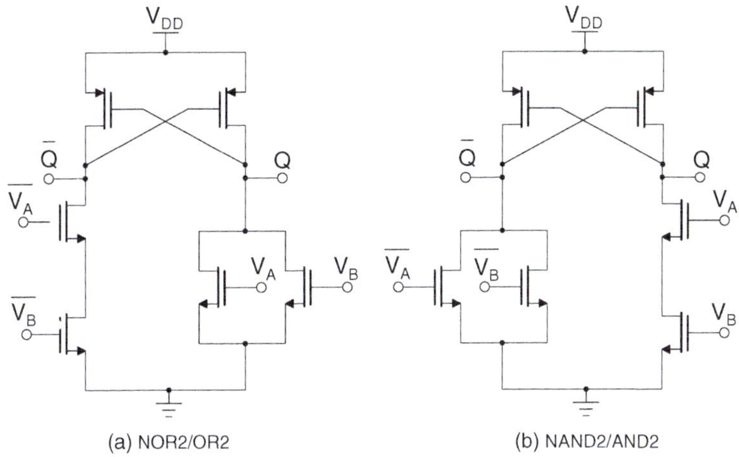

Fig. 6.18 DCVS realization of Boolean gates

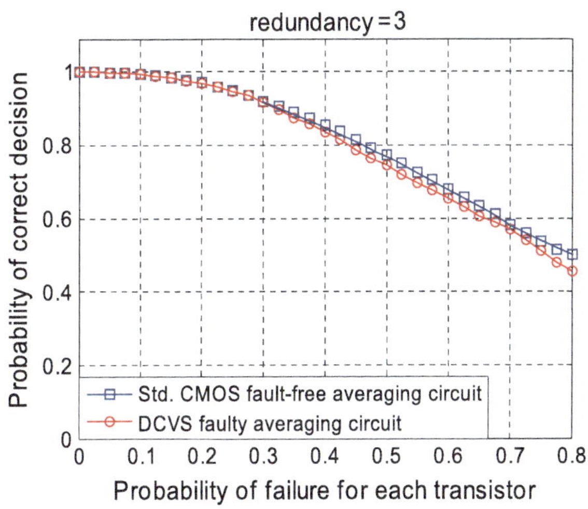

Fig. 6.19 Comparative analysis of the 2-input NAND gate in DCVS and standard CMOS logic with *fault-free* averaging circuit

averaging and thresholding (decision) were not affected by induced faults (this condition is referred in the following as *fault-free averaging circuit*). No significant difference in circuit reliability can be observed under the aforementioned conditions. The reliability of the NOR2 function remains very close to 1 even for device failure rates exceeding 10%.

If faults are induced in the averaging circuit to represent more realistic conditions, however, the overall system reliability drops more rapidly, and the curve does not show saturation for low fault density. Instead, the curve becomes quasi-linear in the critical working range, i.e., for transistor failure probability lower than 30%, as seen in Fig. 6.20 for different redundancy factors. In this case, the use of a differential averaging circuit improves reliability in comparison to standard CMOS, with a significant difference for larger redundancy factors. Moreover, there is no improvement in reliability with respect to redundancy for standard CMOS, which is due to the reliability mainly being dependent on the reliability of the non-redundant averaging circuit.

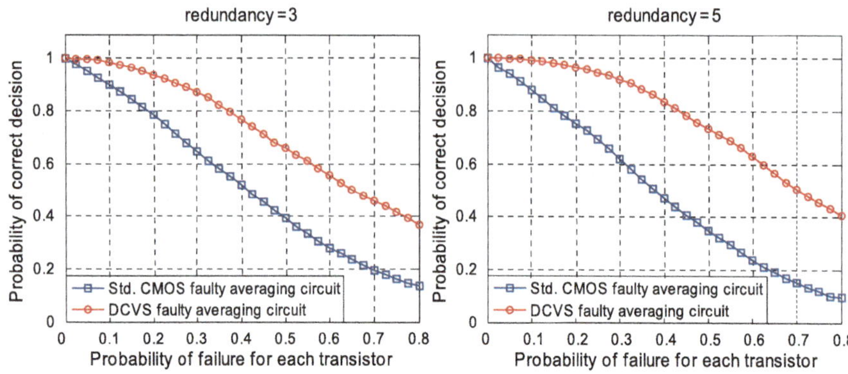

(a) redundancy factor $R = 3$ (b) redundancy factor $R = 5$

Fig. 6.20 Comparative analysis of the 2-input NAND gate in DCVS and standard CMOS logic with a *faulty* averaging circuit, for redundancy of $R = 3$ and $R = 5$

An overall reliability of >99% can be maintained with the DCVS solution under device failure rates of up to 10%, while the reliability of the CMOS solution rapidly drops below 90% for the same device failure rate.

When larger blocks are used as redundant units, the probability of correct operation is reduced with respect to the size of a block. This is depicted in Figs. 6.21 and 6.22 where a gate realizing the complex 4-input function, $f(x1, x2, x3, x4) = x1x4' + (x2x3)' + x1(x2x3)' + x1'x2x3x4$, and a full adder cell are used, respectively, as redundant blocks. Nevertheless, an evident advantage of the DCVS logic realization in comparison to standard CMOS is observed in both cases.

With the 4-input function blocks, there is an improvement in reliability with respect to the increase of redundancy for both configurations (Fig. 6.21). With the full adder block, the advantage of the differential configuration is higher in the output *Sum*, whose path has an increased logic depth and complexity in standard CMOS compared to DCVS. The benefits of the differential architecture become more evident when faults are induced into the averaging circuit (Fig. 6.22a), compared to a fault-free averaging circuit (Fig. 6.22b).

The DCVS logic shows benefits in comparison with standard CMOS logic whenever complex gates or cells (such as a full adders) are used, demonstrating the benefits in terms of reliability of extending differential signaling to the development of more complex digital cells.

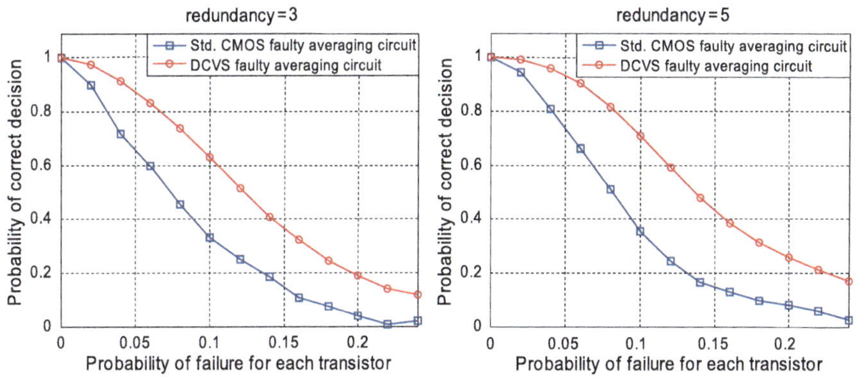

(a) redundancy factor $R = 3$ (b) redundancy factor $R = 5$

Fig. 6.21 Comparative analysis of the 4-input gate realizing the function $f(x1, x2, x3, x4) = x1x4' + (x2x3)' + x1(x2x3)' + x1'x2x3x4$ in DCVS and standard CMOS logic with a *faulty* averaging circuit for redundancy $R = 3$ and $R = 5$

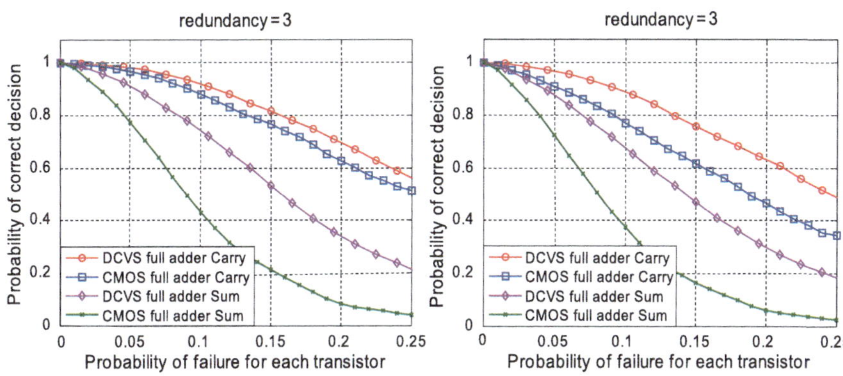

(a) fault-free model of averager circuit (b) faulty model of averager circuit

Fig. 6.22 Comparative analysis of the full adder cell in DCVS and standard CMOS logic for redundancy of $R = 3$ in case of *fault-free* and *faulty* averaging circuit models

6.4 Reliability of SET Systems

Results presented to this point demonstrate clear benefits in using the novel 4LRA fault-tolerant technique compared to RMR for conventional CMOS. However, exploring fault-tolerant techniques' properties when using nanodevices as main logic building blocks is of the highest interest regarding susceptibility of nanodevices to many types of faults.

The novel properties of single-electron (or few-electron) devices appear to offer some interesting and unconventional possibilities that can be exploited for the realization of switching functions. It has already been shown that the input–output characteristics of an inverter can be realized by a simple complementary circuit that is constructed with two SET devices [45–48]. The realization of other elementary

Boolean functions such as OR, NAND, NOR, and a 2-bit adder have also been proposed using the SET devices [7, 48–51]. The ability of the SET to operate with discrete charge levels makes it possible to construct functions that are based on multiple quantized input–output levels, in contrast to the classical Boolean functions operating with two discrete levels.

While offering a number of potential advantages in terms of very high integration density and extremely low power dissipation, SET devices also have serious limitations in terms of output drive capability, speed, and fanout, which would restrict their large-scale integration and interfacing with other system components. Hence, the design of SET-CMOS interface circuits is already gaining importance, as evidenced by some of the proposed hybrid designs in the literature [235–237]. Future systems will likely be based on a hybrid SET-CMOS architecture, where intensive logic or memory functions are performed by very dense, regular arrays of SETs, and the interface functions among blocks are realized in classical, high-speed CMOS components [48, 55, 238].

One of the most significant difficulties of designing complex functions using SETs is the inherent sensitivity of their characteristics to background charge fluctuations. This effect is the result of permanent or transient random variations in local charge due to fabrication irregularities, leakage, or external perturbations such as noise. Background charge effects may permanently or temporarily disrupt device function, rendering one or more SETs inoperative within a functional block in a random manner. To ensure reliable operation and to reduce the sensitivity of devices to background charge effects (especially at room temperature), the device dimensions must be reduced to sub-nanometer levels, which is not very feasible in the foreseeable future. A more likely scenario is that the functional blocks be designed with a certain degree of fine-grained, built-in immunity to permanent and transient faults, such that they are capable of absorbing a number of errors and still be able to perform their functions. A four-layer reliable architecture can be a potential candidate that offers the required level of fault tolerance. In the following sections, the fault-tolerant properties of RMR, AVG with fixed and adjustable threshold, and 4LRA are examined in detail. A possible realization of the averager with adjustable threshold from [239] is shown in Fig. 6.23.

6.4.1 Reliability Evaluation

Simulations of hybrid architectures (SET-CMOS) tend to be slow and complex. Therefore, all designs that have been evaluated are implemented as capacitive input SETs (C-SETs) [237]. The C-SET design is based on the SET inverter [48, 240], which consists of two "complementary" SET transistors (equivalent to nMOS and pMOS transistors), in addition to bias and input capacitors. Manufactured C-SET inverters and multi-input C-SET gates [45, 51] support this approach. Pure C-SET simulation offers significant speed and accuracy advantage over simulations of hybrid architectures (SET-CMOS). The averaging and thresholding operation is

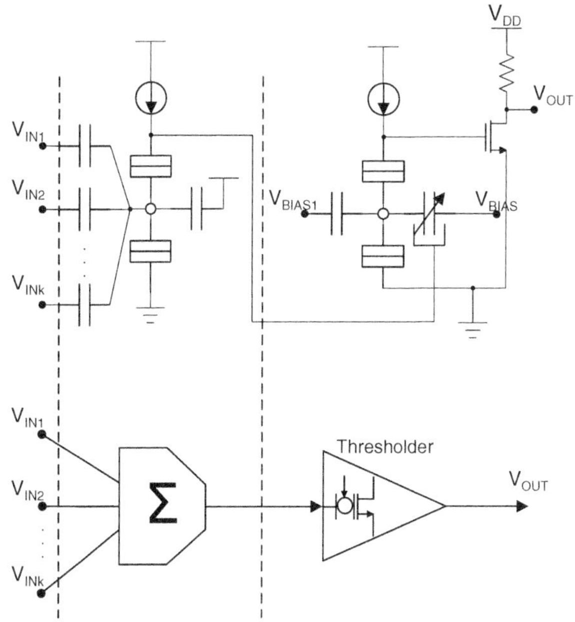

Fig. 6.23 Circuit-level description of the averaging–thresholding hybrid circuit consisting of SET operative circuits driving a MOSFET restoring stage [239]

performed mathematically taking into consideration a hypothetical ideal averager and thresholder, as opposed to a hardware realization shown in Fig. 6.23 which is a SET-CMOS hybrid. The 4LRA using the C-SET realization of 2-input NAND gates with an ideal averager and thresholder is depicted in Fig. 6.24.

The reliability is evaluated using the modified version of the MC tool described in Section 5.3. Instead of using transistor fault model, the modified tool induces geometric variations into SETs by changing the netlists acquired from the schematic. The analysis of the sensitivity to variations is carried out using MATLAB-based modules [241], simultaneously with SIMON [242] (see Fig. 6.25).

Random variations are applied on C-SET elements (capacitors and tunneling junctions). A modified capacitor value is computed from a normal distribution $N(C_0, \sigma_r \cdot C_0)$ centered around nominal value (C_0) and with a relative standard deviation σ_r. The new circuit (with modified capacitors) is subsequently simulated using SIMON, considering all the possible input vectors. The whole procedure including varying the capacitors' values and performing simulations is repeated 10,000 times as a loop in MATLAB, while data are collected in the form of data points, thus forming the transfer function surfaces of the considered block under failure. Subsequently, all simulation results are processed to discriminate among the faulty transfer function surfaces those which can further be thresholded using the fourth layer in order to recover proper circuit behavior. Finally, the related probability of

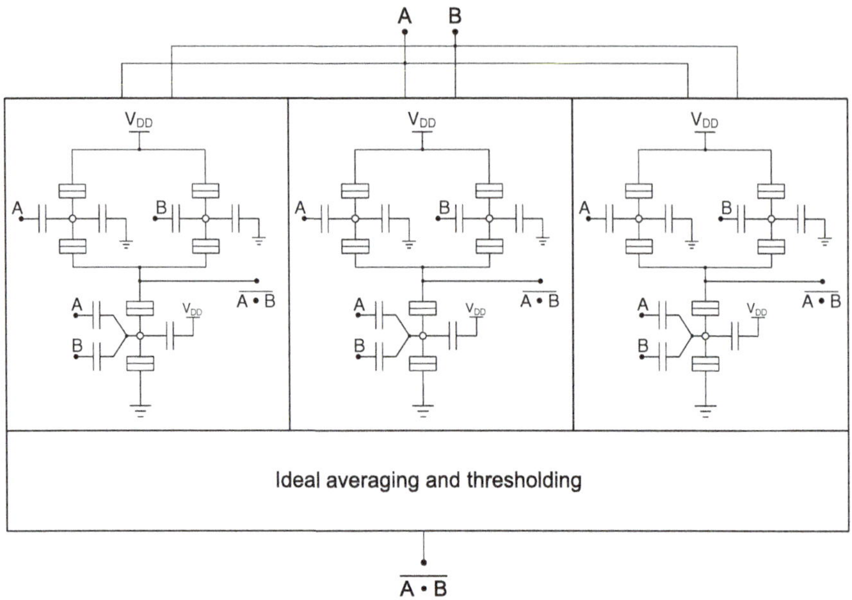

Fig. 6.24 Redundant logic layer with NAND gates as units and ideal averaging and thresholding

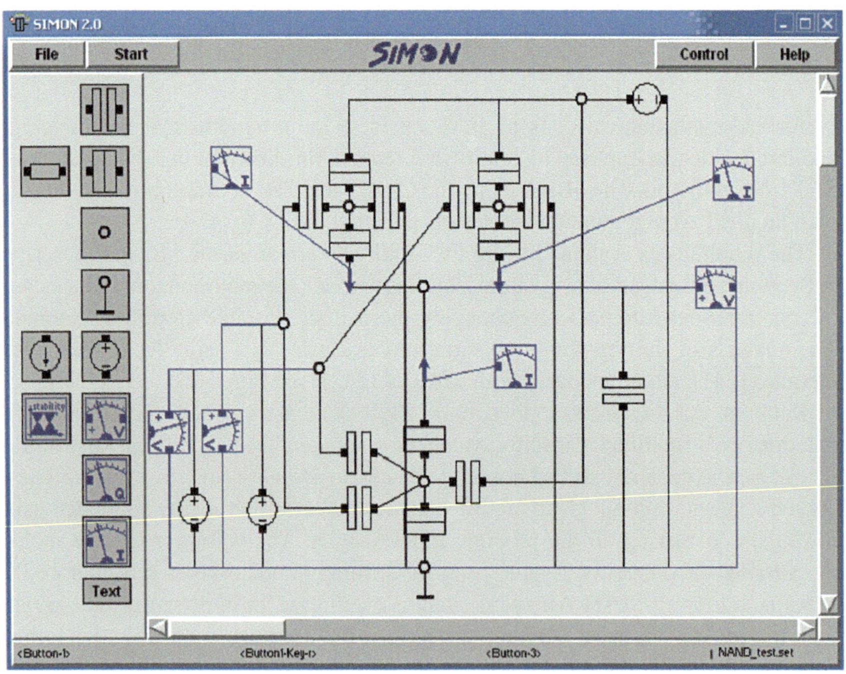

Fig. 6.25 2-input NAND implementation using C-SET technology drawn in SIMON

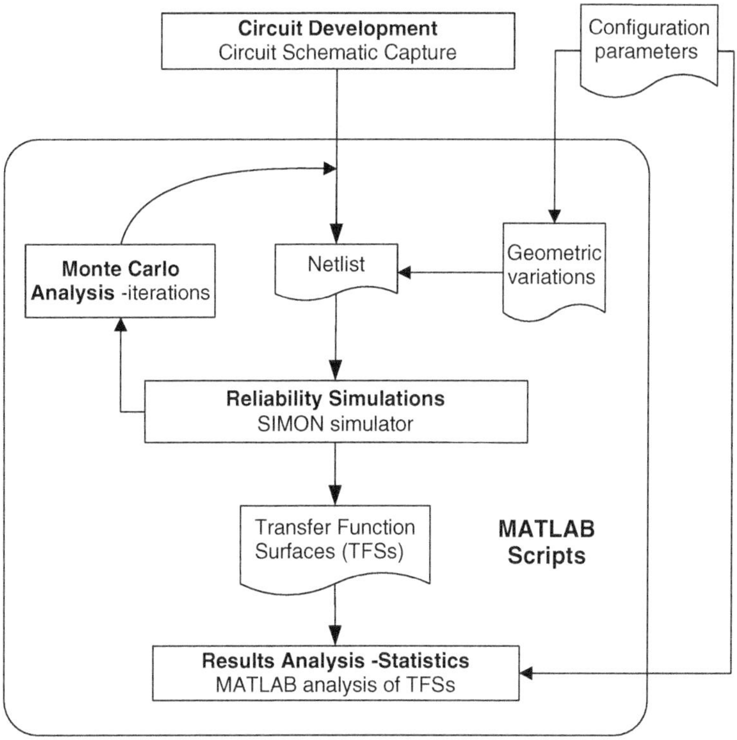

Fig. 6.26 Synthetic flow graph of the tool for SET reliability analysis

correct operation with respect to the probability of fault of a single transistor is calculated. The described steps are depicted in Fig. 6.26.

6.4.2 Comparison of Different Fault-Tolerant Techniques

To compare fault-tolerant techniques, a set of simulations using distinct error densities (variation values) are carried out for different gates, and the reliability of the AVG, AVG-opt, and 4LRA architectures is evaluated using the approach described in the previous section. In all evaluations, a *fault-free* averaging and thresholding unit is assumed.

The gates used for comparison are a 2-input NAND (described in Section 6.4.1) and a full adder (FA). A well-known implementation of the FA using inverting MAJ-3 gates [237, 240] is used (Fig. 6.27). The applied circuit parameters are $C_G = 2$ aF (1 aF for each nSET gate) for NAND and $C_G = 1$ aF for each gate capacitance in the FA, $C_j = 1$ aF for all junction capacitances, $C_b = 5.5$ aF for all bulk capacitances and, C_L=16 aF for the load capacitance. The supply voltage is $V_{DD} = 10$ mV. Simulations are performed at a 1 K operating temperature (as in [240]). The

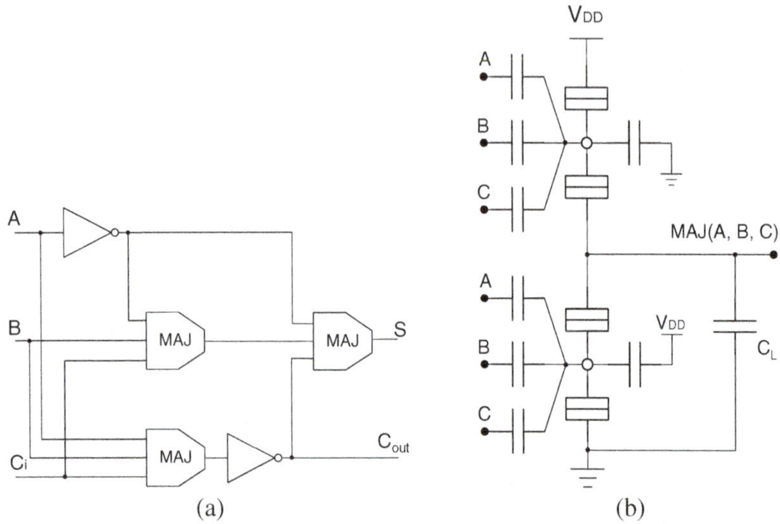

Fig. 6.27 (**a**) MAJ-based SET FA (MAJ-SET); (**b**) MAJ gate based on SET inverter [237, 240]

applied standard deviation of the variability ranges from 1 up to 15%, and differs for different circuits.

In Figs. 6.28, 6.29, and 6.30, the probability failure of different fault-tolerant realizations is plotted vs. the standard deviation of the variability for the NAND gate, the C_{out} output of the FA, and the S output of the FA, respectively.

An advantage of the 4LRA can be observed in terms of the failure probability. The AVG-opt configuration shows significantly better results compared to AVG for

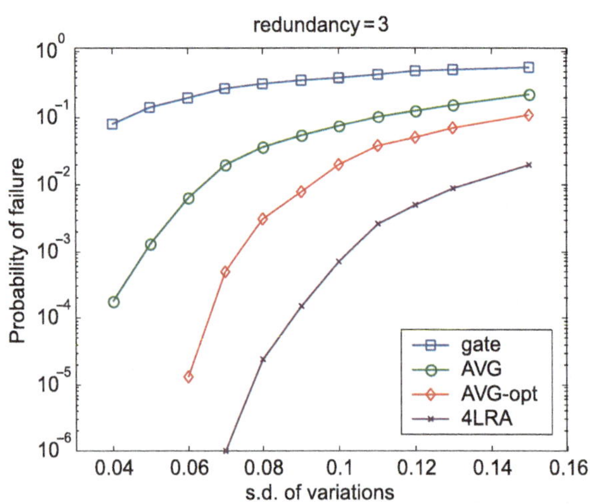

Fig. 6.28 Probability of failure of the NAND gate for different fault-tolerant architectures plotted vs. the standard deviation of variations

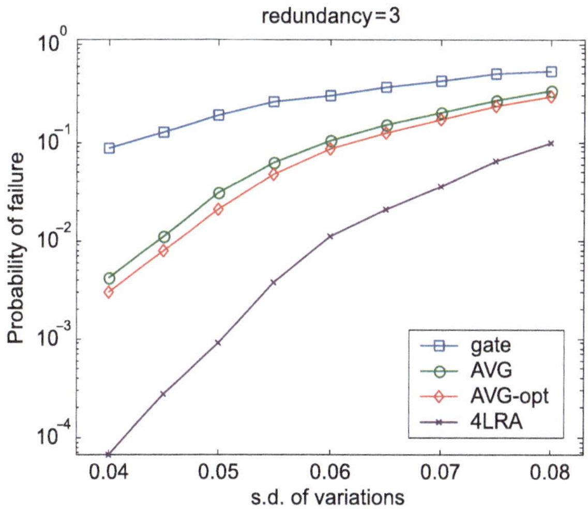

Fig. 6.29 Probability of failure of C_{out} output of the FA for different fault-tolerant architectures plotted vs. the standard deviation of variations

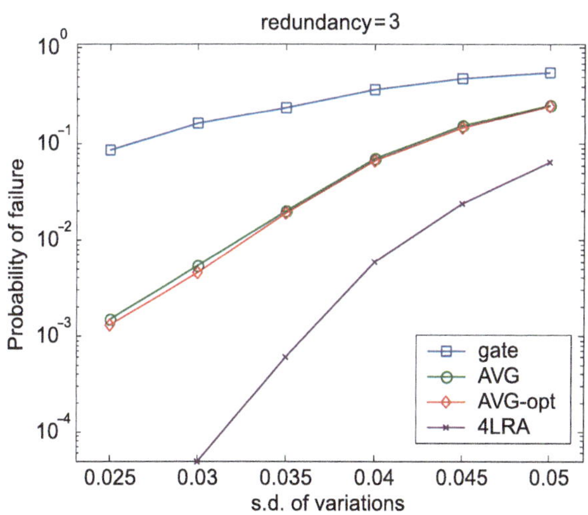

Fig. 6.30 Probability of failure of *S* output of the FA gate for different fault-tolerant architectures plotted vs. the standard deviation of variations

the NAND gate. The fault-tolerant capability of AVG-opt is comparable to 4LRA in this case. The reason for this performance improvement of AVG-opt lies in the fact that output values for logic-0 and logic-1 are not equally probable, and therefore the output value for logic-1 is more sensitive to variations than the output value for logic-0. A low overhead of the AVG-opt realization compared to 4LRA promotes AVG-opt as a better choice. However, when the FA with equally probable output values for logic-0 and logic-1 is used as the main block, the advantage of AVG-opt vs. AVG becomes almost negligible. The overall improvement in reliability of the analyzed fault-tolerant techniques compared to a non-reliable gate is in the range of 100–10,000. Notice also that for the same standard deviation of variations, the probability of failure is increasing and the level of architecture reliability improvement is decreasing for more complex gates (order of complexity: NAND, C_{out} output of the FA, S output of the FA).

Considering the results presented at this point, a conclusion can be drawn, stating that the averaging and thresholding fault-tolerant technique previously evaluated for CMOS circuits also significantly improves the reliability of SET-based designs.

6.5 Summary

Various implementations of the four-layer reliable architecture presented in this chapter demonstrate the versatility of configurations that offer improvements in reliability and yield.

The superiority of 4LRA has been demonstrated over RMR in all presented cases: with fault-free and faulty decision gates, in standard CMOS, and in differential logic at gate level, for individual gates and small circuits. Differential signaling also exhibits superior performance in fault-tolerant architectures compared to single-ended architectures.

4LRA has been applied to circuits built of SET devices as typical representative of nanodevice under research. A specific fault model suitable for SET devices has been implemented and the Monte Carlo tool has been enhanced to support the evaluation of these devices. The analysis shows that the averaging and thresholding fault-tolerant technique can be successfully applied in the process of reliability improvement of inherently unreliable nanodevices such as SETs.

The significance of results presented in this chapter motivates further exploration of the averaging and thresholding techniques in the context of large system reliability evaluation and optimization (Chapter 8).

Chapter 7
Statistical Evaluation of Fault Tolerance Using Probability Density Functions

The precise evaluation of the reliability of logic circuits has a significant importance in highly defective and future nanotechnologies. It allows verifying the theoretical results on the one side and also enables design improvement with respect to their reliability figure by selecting the most suitable (nano)architecture that satisfies all delay, power, area, and reliability requirements on the other.

As a common denominator, most of the methods targeting reliability evaluation (described in Chapter 5) use a single probability value to describe the fault tolerance of each gate in the circuit. This value is the probability of failure of a device (or a logic gate). The benefit of these approaches lies in their relative simplicity of implementation. Analyzing fault-tolerant techniques such as averaging or four-layer architecture, which inherently use analog signals, requires a wide range of output probability values. By analog behavior, a fault-tolerant architecture that operates with analog, continuous values of signals is assumed. The approach prescribing the use of a single probability value to describe the fault tolerance of each device in the system used with averaging and four-layer architectures conducts to inaccurate results.

A method is proposed where the output of a single unit within a so-called redundant layer, i.e., a layer in the fault-tolerant architecture where identical redundant units are present, is described in a statistical manner using probability density functions (PDFs) of the unit output (y). PDFs can be constructed by analyzing the distribution of different faults in the given circuit, as well as the impact of every single fault on the circuit output. PDFs can also be obtained using a Monte Carlo simulator, as the one described in Section 5.3, to acquire output values on a large sample of different fault patterns. The data used in the following section are acquired by applying this approach. PDFs of future nanodevices can be modeled using Gibbs distribution and the approach described in [243, 244]. Finally, PDFs can be modeled using a technique such as the advanced single-pass reliability method described in Section 7.2. The advanced single-pass reliability method represents the modification of the single-pass reliability algorithm [212] to account for permanent errors and to generate output PDFs using the acquired reliability information and PDFs of logic gates.

DOI 10.1007/978-1-4419-6217-1_7,

An alternate approach proposed in [176–178, 245] uses the mean and variance of the output signal to evaluate the reliability of the circuit. A drawback of this method is that the probability of error can only be estimated. One way to estimate the probability of error would consist of applying Chebyshev's inequality [246] to derive an upper bound of the probability of error as in (7.1):

$$P_E \leq \frac{\sigma_y^2}{\sigma_y^2 + \theta^2}, \tag{7.1}$$

where $\theta = |V_{TH} - E\{y\}|$ represents the distance from the expected output level to the gate switching point defined by the gate threshold ($V_{TH} = V_{DD}/2$) and σ_y^2 represents the variance of the output signal. This inequality is valid for any PDF, but it gives an upper bound that is too large in most cases of interest. Moreover, it will be demonstrated in the following section that two different PDFs with equal mean and variance value can yield a completely different probability of error.

7.1 Statistical Method for the Analysis of Fault-Tolerant Techniques

A statistical method that provides the probability of error of circuits utilizing different fault-tolerant techniques is presented and verified, using data obtained by means of MC simulations. The output PDF of various decision gates used in conjunction with R redundant units is obtained using mathematical transformations of the unit output PDFs. Finally, the probability of error is derived from the output PDF of the decision gate. The analyzed fault-tolerant techniques employ the following decision gates:

- Majority voter (MV) which guarantees correct operation even when $\frac{R-1}{2}$ out of R redundant units are failing.
- Averager with optimal fixed threshold (AVG), where a fixed threshold is set in order to minimize the probability of failure for logic-0 and logic-1, simultaneously.
- Averager with adjustable threshold (4LRA) that together with redundant units forms the full four-layer reliable architecture.

PDFs are constructed by analyzing circuit outputs obtained from MC simulations for each input vector, applying numerous fault patterns. A circuit with n inputs has 2^n different input vectors and 2^n corresponding output values. Let $Y_i, i \in \{1, \ldots, 2^n\}$, be random variables that correspond to output values of a circuit. PDFs that correspond to those random variables are marked with $h_i, i \in \{1, \ldots, 2^n\}$. These variables can be divided into two sets, H_1 consisting of output values corresponding to input vectors that produce a logic-1 output and H_0 consisting of output values corresponding to input vectors that produce a logic-0 output. Let $Y_{\min 1}$ be a random variable that corresponds to $y_{\min} = \min H_1$ and $Y_{\max 0}$ be a random variable that corresponds to $y_{\max} = \max H_0$. Two additional PDFs of interest are PDFs that correspond to random variables $Y_{\min 1}$ and $Y_{\max 0}$, named worst-case logic-1 and worst-case logic-0 in further explanations (marked as $h_{\min 1}$ and $h_{\max 0}$, respectively).

These PDFs are continuous, since the output voltage of a faulty circuit may potentially take any value, and is only (softly) limited by the power rails. The individual values of the parameters used in the transistor-level fault model for MC simulations of circuits obey a normal distribution, and therefore the number of different possible output values is unlimited. In order to maintain the generality of the approach, the mathematical apparatus presented hereafter will also use continuous functions. On the other hand, the actual calculations implemented in MATLAB with custom scripts are performed on discrete data sets.

In the following example, the worst-case logic-0 and logic-1 output PDFs for two different circuits (namely $h_{\min 1,a}$ and $h_{\min 1,b}$) are evaluated; $h_{\min 1,a}$ and $h_{\min 1,b}$ are shown in Fig. 7.1a, b. Values located along the x-axis are plotted in relative units of $V_{\rm DD}$. PDFs are continuous and the area under PDFs is equal to 1. This example is intentionally selected as very unfavorable in order to show the level of inaccuracy in extreme cases, when only mean and variance are used in the evaluation process. Still, the hypothetical output PDF is similar to a realistic gate PDF.

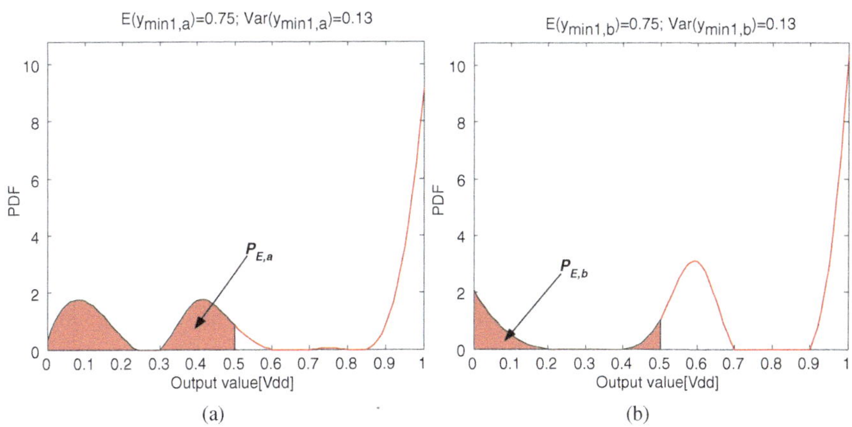

Fig. 7.1 PDF of the unit output for the worst-case logic-1, mean and variance: (**a**) $h_{\min 1,a}$, $E\{y_{\min 1,a}\} = 0.75$, $\sigma^2\{y_{\min 1,a}\} = 0.13$; (**b**) $h_{\min 1,b}$, $E\{y_{\min 1,b}\} = 0.75$, $\sigma^2\{y_{\min 1,b}\} = 0.13$

Considering that only outputs that are above the threshold ($V_{\rm TH} = V_{\rm DD}/2$) are correct, the probability of error for both circuits (also depicted in 7.1) with the expected output at logic-1 is expressed as follows (7.2):

$$P_{{\rm E},a/b} = \int_0^{0.5} h_{\min 1,a/b}(x)dx \tag{7.2}$$

Both PDFs $P_{E,a}$ and $P_{E,b}$ have the same mean and variance ($E\{y_{\min 1,a}\} = E\{y_{\min 1,b}\} = 0.75$ and $\sigma^2\{y_{\min 1,a}\} = \sigma^2\{y_{\min 1,b}\} = 0.13$) but when evaluated using (7.2), the probability of error is 0.335 in the first case (Fig. 7.1a) and 0.211 in the second (Fig. 7.1b). Using Chebyshev's inequality (7.1) yields a probability of error equal to 0.675. Such a large discrepancy suggests that in some cases, the mean and variance parameters are insufficient to obtain an accurate evaluation, regardless of the method that is used for the evaluation.

The redundant unit circuit, whose output PDFs are used as examples in calculations of the statistical method, is a small circuit realized using 2-input NAND gates depicted in Fig. 7.2. The exact nature of the unit circuit is irrelevant in terms of the derived method and calculation complexity.

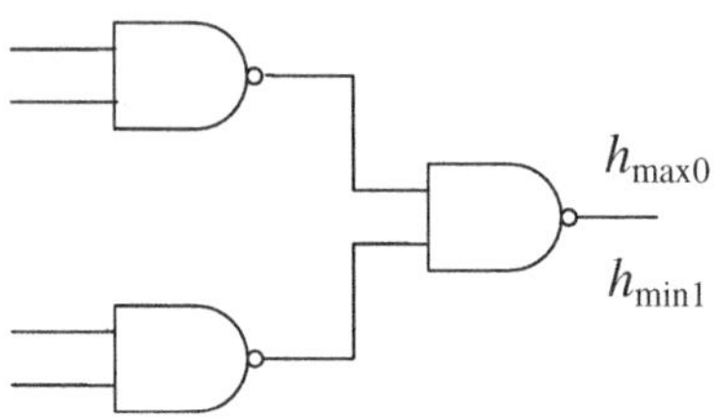

Fig. 7.2 Simple circuit example realized with 2-input NAND gates used as a logic unit

The worst-case logic-0 and logic-1 PDFs for the analyzed unit are depicted in Fig. 7.3a, b. According to the aforementioned definition, the worst-case logic-0 represents the highest value of the output in the presence of variations, which is expected to be a logic-0 level in the absence of variations (and accordingly for the worst-case logic-1). Values located along the x-axis are plotted in relative units of V_{DD} in Figs. 7.3, 7.4, and 7.5. The plotted PDF values are continuous and interpolated using a 100-point histogram. The area under the PDFs is equal to 1. The non-monotonic nature of the curves and pronounced peaks is due to the fact that some circuit states are more common than others.

The probabilities of faulty output of the unit for logic-0 and logic-1 (Fig. 7.3a, b), respectively, are given as

$$P_{E0} = \int_{0.5}^{1} h_{\max 0}(x)dx$$

and (7.3)

$$P_{E1} = \int_{0}^{0.5} h_{\min 1}(x)dx$$

where PDFs for worst-case logic-0 and worst-case logic-1 are marked with $h_{\min 1}$ and $h_{\max 0}$, respectively. The threshold which determines the correctness of the unit operation is assumed to be set to $V_{DD}/2$. The random variables which follow these PDFs are assumed to be correlated and P_{E01} is defined as the probability of simultaneous faulty output of the circuit for logic-0 and logic-1, i.e.,

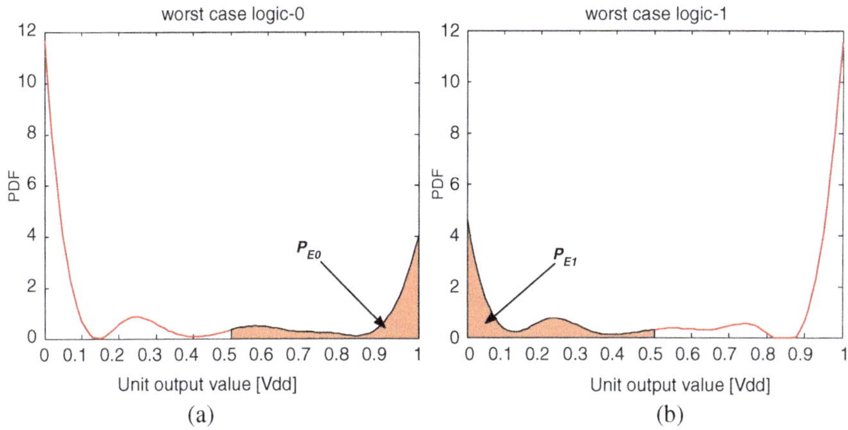

Fig. 7.3 PDF of unit output for (**a**) the worst-case logic-0 (h_0); (**b**) the worst-case logic-1 (h_1)

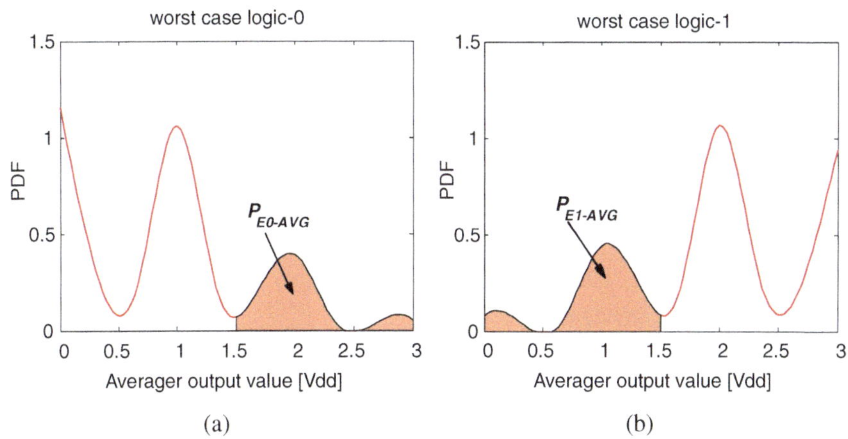

Fig. 7.4 PDF of averager output for (**a**) worst-case logic-0 (h_0^{*3}); (**b**) worst-case logic-1 (h_1^{*3})

$$P_{\mathrm{E01}} = \Pr(Y_{\max 0} > V_{\mathrm{DD}}/2 \text{ and } Y_{\min 1} < V_{\mathrm{DD}}/2). \tag{7.4}$$

The total probability of the unit failure ($P_{\mathrm{E\_unit}}$) is given as

$$P_{\mathrm{E\_unit}} = P_{\mathrm{E0}} + P_{\mathrm{E1}} - P_{\mathrm{E01}}, \tag{7.5}$$

where P_{E0} and P_{E1} are taken from (7.3). P_{E01} can be acquired using the modified single-pass reliability tool as explained in Section 7.2.1.

The probability of failure for a logic-0 (logic-1) of three redundant units with a majority voter ($P_{\mathrm{E0/1\_MV3}}$) is given as

$$P_{\mathrm{E0/1\_MV3}} = 3P_{\mathrm{E0/1}}^2(1 - P_{\mathrm{E0/1}}) + P_{\mathrm{E0/1}}^3. \tag{7.6}$$

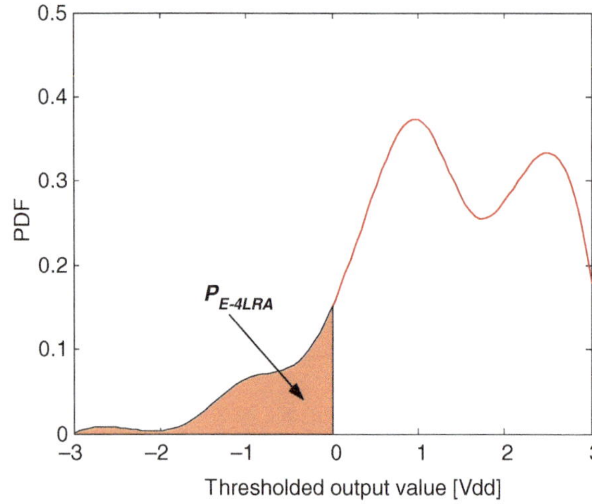

Fig. 7.5 PDF of 4LRA output (h_{TH})

The total probability of failure of three redundant units with a majority voter (P_{E_MV3}) is given as

$$P_{E\_MV3} = P_{E0\_MV-3} + P_{E1\_MV-3} - P_{E01\_MV3}, \tag{7.7}$$

where P_{E01_MV3} is the probability of simultaneous faulty output of the unit for logic-0 and logic-1 and can be further defined as

$$\begin{aligned} P_{E01\_MV3} = 3 \cdot 2 \cdot P_{E01}(1 - P_{E01})^2(P_{E0} - P_{E01})(P_{E1} - P_{E01}) \\ + 3 \cdot P_{E01}^2(1 - P_{E01}) + P_{E01}^3. \end{aligned} \tag{7.8}$$

In a general case of R redundant units with a majority voter (MV), the probability of failure for logic-0 (logic-1) ($P_{E0/1\_MV}$) is given as

$$P_{E0/1\_MV} = \sum_{i=\frac{R+1}{2}}^{R} \binom{R}{i} P_{E0/1}^{i}(1 - P_{E0/1})^{R-i}. \tag{7.9}$$

The total probability of failure of the MV is given as

$$P_{E\_MV} = P_{E0\_MV} + P_{E1\_MV} - P_{E01\_MV}, \tag{7.10}$$

where P_{E01_MV} is the probability of simultaneous faulty output of the majority voter for logic-0 and logic-1 and can further be derived as the sum of all combinations when at least $\frac{R+1}{2}$ outputs of the majority voter for logic-0 and logic-1 fail simultaneously:

$$P_{\mathrm{E01\_MV}} = \sum_{i=\frac{R+1}{2}}^{R} \binom{R}{i} P_{\mathrm{E01}}^{i}(1-P_{\mathrm{E01}})^{R-i} + \sum_{i=1}^{\frac{R-1}{2}} \binom{R}{i} P_{\mathrm{E01}}^{i}(1-P_{\mathrm{E01}})^{R-i} \cdot \binom{R-i}{\frac{R+1}{2}-i}\binom{\frac{R-1}{2}}{\frac{R+1}{2}-i}[(P_{\mathrm{E0}}-P_{\mathrm{E01}})(P_{\mathrm{E1}}-P_{\mathrm{E01}})]^{\frac{R+1}{2}-i}. \tag{7.11}$$

The expression of the fault tolerance of AVG and 4LRA is derived in the following by transforming PDFs in the averaging and thresholding layer. Random variables Y_i, $i \in \{1,\ldots,2^n\}$, that correspond to output values of R redundant units are summed using the averager circuit forming new random variables $Z_i = RY_i$, $i \in \{1,\ldots,2^n\}$. These variables can be divided into two sets, $H_1^{(R)}$ consisting of output values corresponding to input vectors that produce a logic-1 output and $H_0^{(R)}$ consisting of output values corresponding to input vectors that produce a logic-0 output. Let $Y_{\min 1}^{(R)}$ be a random variable that corresponds to $y_{\min}^{(R)} = \min H_1^{(R)}$ and $Y_{\max 0}^{(R)}$ be a random variable that corresponds to $y_{\max}^{(R)} = \max H_0^{(R)}$. Two additional PDFs of interest are PDFs that correspond to random variables $Y_{\min 1}^{(R)}$ and $Y_{\max 0}^{(R)}$ and are marked as $h_{\min 1}^{(R)}$ and $h_{\max 0}^{(R)}$, respectively. Here, the approximation that $i \in \{1,\ldots,2^n\}$ for which $Y_{\min 1}^{(R)} = Z_i$ and $Y_{\min 1}{=}Y_i$ and $j \in \{1,\ldots,2^n\}$ for which $Y_{\max 0}^{(R)} = Z_j$ and $Y_{\max 0} = Y_j$) is always valid. In other words, an input vector that simultaneously produces the worst-case output in each redundant unit exists for each combination of faulty patterns within units. Since this is not fulfilled in all the cases, there are faulty patterns for which the actual worst-case logic-0 (logic-1) value is smaller (larger) than the value given by $Y_{\max 0}^{(R)}$ ($Y_{\min 1}^{(R)}$). Taking this into consideration, PDFs defined as $h_{\max 0}^{(R)}$ and $h_{\min 1}^{(R)}$ represent the worst case with respect to probability of failure at the output of the averager.

Following the introduced approximation, PDFs for the worst-case logic-0 and logic-1 considered after the averaging operation ($h_{\max 0}^{(R)}$ and $h_{\min 1}^{(R)}$) become PDFs of a sum of R identical and independent random variables, since errors in each redundant unit are uncorrelated, i.e., common-mode (or common cause) failures are not present in the redundant system [247], which is represented by the R-fold convolution [246] in (7.12):

$$h_{\max 0}^{(R)} \approx \underbrace{h_{\max 0} * h_{\max 0} * \cdots * h_{\max 0}}_{R}$$

$$\approx \underbrace{\int_{-\infty}^{\infty} \cdots \int_{-\infty}^{\infty}}_{R-1} h_{\max 0}(x_1) h_{\max 0}(x_2 - x_1)dx_1 \cdots h_{\max 0}(x_R - x_{R-1})dx_{R-1}$$

$$h_{\min 1}^{(R)} \approx \underbrace{h_{\min 1} * h_{\min 1} * \cdots * h_{\min 1}}_{R}$$
$$\approx \underbrace{\int_{-\infty}^{\infty} \cdots \int_{-\infty}^{\infty}}_{R-1} h_{\min 1}(x_1) h_{\min 1}(x_2 - x_1) dx_1 \cdots h_{\min 1}(x_R - x_{R-1}) dx_{R-1}. \tag{7.12}$$

In order to simplify the calculation without losing generality, an averager which is not performing rescaling of the output is considered and its output value range remains between 0 and RV_{DD} (averager with R inputs ranging between 0 and V_{DD}). The corresponding PDFs for logic-0 and logic-1 for averaging of three redundant units are represented in Fig. 7.4a, b. Due to the convolution operation, and the non-monotonicity (existence of peaks) in the initial PDFs, some pronounced local maxima could be observed. The R-fold convolution for high redundancy factors (in practice $R > 20$) converges to a normal distribution with the same mean and the variance that is R times smaller than the initial PDF's variance (7.13), according to the central limit theorem [246]

$$E\left(\frac{1}{R}\sum_{i=1}^{R} X_i\right) = \mu, \quad \mathrm{Var}\left(\frac{1}{R}\sum_{i=1}^{R} X_i\right) = \frac{\sigma^2}{R}. \tag{7.13}$$

In (7.13), $X_1, \ldots, X_n$ represent independent random variables with a mean μ, and a variance σ^2, and whose PDFs are $h_{\max 0}$ and $h_{\min 1}$. Similar expression and results are also presented in the work of Martorell et al. [178, 245].

Taking into account that only output values above the threshold (V_{TH}) are correct for logic-1 and below threshold for logic-0,

$$P_{\mathrm{E0\_AVG}} = \int_{V_{\mathrm{TH}}}^{R} h_{\max 0}^{(R)}(t)dt$$
$$\text{and} \tag{7.14}$$
$$P_{\mathrm{E1\_AVG}} = \int_{0}^{V_{\mathrm{TH}}} h_{\min 1}^{(R)}(t)dt$$

are marked as the probabilities of a faulty output of the averager for logic-0 and logic-1 (Fig. 7.4a, b), respectively. Numerical values obtained from (7.14) are used in (7.15) to acquire the probability of failure of the averager ($P_{\mathrm{E\_AVG}}$),

$$P_{\mathrm{E\_AVG}} = P_{\mathrm{E0\_AVG}} + P_{\mathrm{E1\_AVG}} - P_{\mathrm{E01\_AVG}}, \tag{7.15}$$

where $P_{\mathrm{E01\_AVG}}$ represents the probability of simultaneous faulty output of AVG for logic-0 and logic-1. Accurately determining $P_{\mathrm{E01\_AVG}}$ requires the knowledge of the conditional PDF for logic-0 when the output of AVG for logic-1 is faulty and the conditional PDF for logic-1 when the output of AVG for logic-0 is faulty. The analytical derivation of these conditional probabilities is an intricate task. Therefore, an approximation

$$P_{\mathrm{E01\_AVG}} \approx P_{\mathrm{E01\_MV}} \tag{7.16}$$

is used and justified by further comparison with simulated results.

When $P_{\mathrm{E0\_AVG}}$ is significantly larger than $P_{\mathrm{E1\_AVG}}$ (or vice versa) for the default threshold value of half of the output range ($V_{\mathrm{TH}} = RV_{\mathrm{DD}}/2$), $P_{\mathrm{E\_AVG}}$ can be reduced by setting the threshold to the optimal value (V_{opt}) defined as the numerical solution of the equation $dP_{\mathrm{E\_AVG}}/dV_{\mathrm{TH}} = 0$.

The threshold in the fourth layer has been considered to be fixed and set to the optimal value that provides the minimum probability of failure ($P_{\mathrm{E\_AVG}}$), i.e., $V_{\mathrm{TH}} = V_{\mathrm{opt}}$, where V_{opt} is taken as the numerical solution of the equation $dP_{\mathrm{E\_AVG}}/dV_{\mathrm{TH}} = 0$. This is a predefined value set for each AVG fault-tolerant unit in the design phase.

An adaptable threshold in 4LRA can correct the output of the averaging layer if the worst-case value for logic-1 at the output of the averaging layer is higher than the worst-case logic-0, hence the difference between the random variables for the worst-case logic-1 and logic-0 (whose PDFs are $h^{(R)}_{\min 1}$ and $h^{(R)}_{\max 0}$) is positive. The PDF of a difference of two random variables is given in (7.17) and depicted in Fig. 7.5 (for three redundant units) [246]:

$$h_{\mathrm{TH}}(t) = h^{(R)}_{\min 1}(t) * h^{(R)}_{\max 0}(-t) = \int_{-\infty}^{\infty} h^{(R)}_{\min 1}(x) h^{(R)}_{\max 0}(x-t)dx. \tag{7.17}$$

h_{TH} exists in the range $[-RV_{\mathrm{DD}}, RV_{\mathrm{DD}}]$ and the probability of failure of 4LRA ($P_{\mathrm{E\_4LRA}}$) only takes the values of h_{TH} for positive differences (in the range $[0, RV_{\mathrm{DD}}]$). Equation (7.17) assumes that random variables that correspond to PDFs $h^{(R)}_{\min 1}$ and $h^{(R)}_{\max 0}$ are independent. Since this is not fulfilled, a correction factor has to be included in the expression of $P_{\mathrm{E\_4LRA}}$ to account for the cases of simultaneous failure for logic-0 and logic-1. This correction factor is equal to $P_{\mathrm{E01\_AVG}} \approx P_{\mathrm{E01\_MV}}$ since 4LRA fails whenever the averager output values for logic-0 and logic-1 are simultaneously faulty. Finally, the expression of $P_{\mathrm{E\_4LRA}}$, also illustrated in Fig. 7.5, is

$$P_{\mathrm{E\_4LRA}} \approx \int_{-R}^{0} h_{\mathrm{TH}}(t)dt + P_{\mathrm{E01\_MV}}. \tag{7.18}$$

Figures 7.3, 7.4, and 7.5 are plotted using PDFs acquired from the MC tool. An example circuit (depicted in Fig. 7.2) is used as a unit in the logic layer. The number of MC iterations is set to 320,000 in order to minimize the error due to sampling as explained in Section 5.3. The probability of fault per transistor (p_{f}), i.e., the probability that in each MC iteration a transistor fault model is applied to each individual transistor, is 10%. Such a high value is chosen for the purpose of easier illustration in the figures. A set of MATLAB scripts have been developed to automate the described process related to the analytical evaluation of reliability, from PDFs acquired using the MC tool. Applying (7.14), (7.15), (7.16), (7.17), and (7.18), probabilities of failure for different decision gates, different defect densities, and different redundancy factors are calculated and summarized in Table 7.1 (denoted as "calculated"). Also, results from MC simulations are evaluated in each iteration, directly applying averaging and thresholding and are given for the purpose of verification (denoted as "simulated" in Table 7.1). Small values (<5%) of the relative error between calculated and simulated results for $P_{\mathrm{E01\_AVG}}$ confirm the approximation introduced with (7.16). Moreover, small values of the relative error between calculated and simulated results for $P_{\mathrm{E\_4LRA}}$ confirm the validity of (7.18) and justify the use of the correction factor.

A novel statistical method for the analysis of fault-tolerant (MV, AVG, and 4LRA) techniques is presented. The method enables accurate evaluation (relative error smaller than 5%) of the reliability (probability of failure) when PDFs of the unit output are known. The method is implemented in MATLAB scripts and the

Table 7.1 Probabilities of error (P_{E}) for different fault-tolerant techniques, different defect densities (p_{f}) and different redundancy factors (a) $R = 3$, (b) $R = 5$, and (c) $R = 7$

(a)

$R = 3$	$p_{\mathrm{f}}(\%)$	$P_{\mathrm{E\_unit}}$	$P_{\mathrm{E\_MV}}$	$P_{\mathrm{E01\_AVG}}$	$P_{\mathrm{E\_AVG}}$	$P_{\mathrm{E\_4LRA}}$
Calculated	0.2	0.01323	2.765×10^{-4}	3.641×10^{-7}	2.377×10^{-5}	2.444×10^{-6}
Simulated		0.01323	2.771×10^{-4}	3.541×10^{-7}	2.398×10^{-5}	2.451×10^{-6}
Calculated	1	0.0589	5.925×10^{-3}	7.748×10^{-5}	5.064×10^{-3}	2.711×10^{-4}
Simulated		0.0589	5.915×10^{-3}	7.583×10^{-5}	4.997×10^{-3}	2.616×10^{-4}
Calculated	5	0.2553	0.1196	7.250×10^{-3}	0.0968	2.314×10^{-2}
Simulated		0.2553	0.1187	7.174×10^{-3}	0.0963	2.299×10^{-2}

(b)

$R = 5$	$p_{\mathrm{f}}(\%)$	$P_{\mathrm{E\_gate}}$	$P_{\mathrm{E\_MV}}$	$P_{\mathrm{E01\_AVG}}$	$P_{\mathrm{E\_AVG}}$	$P_{\mathrm{E\_4LRA}}$
Calculated	0.2	0.01323	6.233×10^{-6}	5.342×10^{-10}	4.804×10^{-6}	1.196×10^{-9}
Simulated		0.01323	6.221×10^{-6}	5.264×10^{-10}	4.696×10^{-6}	1.149×10^{-9}
Calculated	1	0.0589	6.223×10^{-4}	1.644×10^{-6}	4.731×10^{-4}	3.003×10^{-6}
Simulated		0.0589	6.221×10^{-4}	1.607×10^{-6}	4.736×10^{-4}	2.930×10^{-6}
Calculated	5	0.2553	5.047×10^{-2}	1.614×10^{-3}	3.949×10^{-2}	4.084×10^{-3}
Simulated		0.2553	5.041×10^{-2}	1.589×10^{-3}	3.917×10^{-2}	3.903×10^{-3}

(c)

$R = 7$	$p_{\mathrm{f}}(\%)$	$P_{\mathrm{E\_gate}}$	$P_{\mathrm{E\_MV}}$	$P_{\mathrm{E01\_AVG}}$	$P_{\mathrm{E\_AVG}}$	$P_{\mathrm{E\_4LRA}}$
Calculated	0.2	0.01323	1.474×10^{-7}	8.172×10^{-13}	9.967×10^{-8}	1.048×10^{-12}
Simulated		0.01323	1.481×10^{-7}	7.988×10^{-13}	1.007×10^{-7}	1.003×10^{-12}
Calculated	1	0.0589	6.787×10^{-5}	3.780×10^{-8}	4.497×10^{-5}	4.835×10^{-8}
Simulated		0.0589	6.753×10^{-5}	3.595×10^{-8}	4.416×10^{-5}	4.688×10^{-8}
Calculated	5	0.2553	2.329×10^{-2}	3.832×10^{-4}	1.627×10^{-2}	8.115×10^{-4}
Simulated		0.2553	2.315×10^{-2}	3.632×10^{-4}	1.591×10^{-2}	7.938×10^{-4}

implementation is very fast (all calculations presented in Table 7.1 are performed in less than 100 ms). A prerequisite for the efficient use of the method is a fast and accurate generation of output PDFs for arbitrary unit (any circuit type or size) which is a topic of the following section.

7.2 Advanced Single-Pass Reliability Evaluation Method

A fast and accurate reliability evaluation tool is necessary in order to acquire output PDFs of a logic unit of an arbitrary size. From all the methods for the reliability analysis presented in Chapter 5, analytical and experimental methods (discrete-event simulation) are directly applicable to circuit output PDF generation. An example of discrete-event simulation is a Monte Carlo framework such as our software tool described in Section 5.3 which uses fault injection and simulation. Although parallelizable and scalable, MC tools are still not efficient to use with large circuits.

Analytical methods for the reliability analysis (as also mentioned in Section 5.3) are applicable to very simple structures such as 2-input and 3-input gates and regular fabrics [13, 123]. Despite the fact that they can be applied to large multi-level circuits, a significant loss in accuracy is observed due to simplified assumptions and compositional rules.

Numerical methods enabling reliability evaluation use a single probability value to describe the fault tolerance of each gate in the circuit. Moreover, the output result of the evaluation is a set of probability values for some input vector sets. Recent presented advances in reliability analysis such as probabilistic transfer matrices (PTMs) [190] and Bayesian networks [207] require significant runtimes for small benchmarks. Other works like [210] do not satisfy accuracy requirements. Two most recent approaches, single-pass reliability analysis [248] and signal probability analysis [214] satisfy accuracy and scalability requirements. However, as shown in Table 5.1, the single-pass reliability analysis tool offers better performance in terms of speed and scalability and requires less memory resources. Moreover, the output of the single-pass reliability tool (probabilities of failure for logic-0 and logic-1) is more suitable for the worst-case logic-0 and logic-1 PDF modeling than values of the probability of failure for each input vector supplied by the signal probability analysis tool.

The single-pass reliability evaluation tool [212] presents an implementation of a fast, accurate, and scalable novel algorithm for reliability analysis. The original algorithm is intended for transient errors which can be modeled as a symmetrical flip (from $0 \rightarrow 1$ or $1 \rightarrow 0$) of a gate output, with the same probability of error.

The modified algorithm, also covering permanent errors with the different values of gate probability of error for the worst-case logic-0 and logic-1, is presented. Transient errors (whose effects are observable no longer than one clock period) can affect the output value that corresponds to *only one* input vector active at that time. Permanent errors on the other hand are present all the time and affect the output value that corresponds to *each* input vector. The algorithm is also extended to evaluate the joint probability of simultaneous faulty output of the circuit for logic-0 and logic-1 which is necessary factor used in the statistical method presented in the previous section.

The advanced single-pass reliability evaluation method consists of two steps. (i) In the first step the modified single-pass reliability evaluation tool is used to evaluate the probability of error for the worst-case logic-0 and logic-1 of the arbitrary circuit outputs. (ii) In the second step, using this probability data and PDFs for individual gates, the circuit output PDFs are modeled. The accuracy of the presented method is also demonstrated.

7.2.1 Modified Single-Pass Reliability Evaluation Tool

7.2.1.1 Single-Pass Reliability Algorithm

In the original algorithm [248], gates are topologically sorted and processed in a single pass from the inputs to the outputs. Topological sorting ensures that before a gate is processed, the effects of multiple gate failures in the fanin cone of the gate are computed and stored at the inputs of the gate. At the core of this algorithm is the observation that an error at the output of any gate results from the cumulative effect of a local error component attributed to the probability of failure of the observed gate and a propagated error component attributed to the failure of gates in its fanin cone. The effect of reconvergent fanout on error probabilities is addressed using correlation coefficients.

An error at the output of any gate results from the cumulative effect of a local error component attributed to the probability of failure of the observed gate and a propagated error component attributed to the failure of gates in its fanin cone. The following two events are of importance:

- $0 \rightarrow 1$, which marks an event where the output of the gate is at logic-1, whereas its fault-free value is logic-0 (referred to as worst-case logic-0 error);
- accordingly, $1 \rightarrow 0$, which marks an event where the output of the gate is at logic-0, whereas its fault-free value is logic-1 (referred to as worst-case logic-1 error).

The equivalent to those two events in the analog domain (when using analog signals) are the worst-case logic-0 and logic-1 faults introduced in the previous section.

Following symbol convention is established, covering the analytical developments presented in this chapter:

- the arbitrary gate output is marked as g;
- the total error probability at g of a $0 \rightarrow 1$ event is marked as $\Pr(g_{0\rightarrow 1})$;
- the total error probability at g of a $1 \rightarrow 0$ event is marked as $\Pr(g_{1\rightarrow 0})$;
- the joint total propagation error probability at g of $0 \rightarrow 1$ and $1 \rightarrow 0$ events is marked as $\Pr(g_{\text{joint}})$;

- the propagation error probability at g (which is defined further in the text) of a $0 \rightarrow 1$ event is marked as $P_{\text{p}}(g_{0\rightarrow 1})$;
- the propagation error probability at g of a $1 \rightarrow 0$ event is marked as $P_{\text{p}}(g_{1\rightarrow 0})$;
- the joint propagation error probability at g of $0 \rightarrow 1$ and $1 \rightarrow 0$ events is marked as $P_{\text{p}}(g_{\text{joint}})$;
- the single gate error probability of a $0 \rightarrow 1$ event is marked as $P_{\epsilon 0}$;
- the single gate error probability of a $1 \rightarrow 0$ event is marked as $P_{\epsilon 1}$;
- the joint single gate error probability of $0 \rightarrow 1$ and $1 \rightarrow 0$ events is marked as $P_{\epsilon 01}$.

In general, $\Pr(g_{0\rightarrow 1}) \neq \Pr(g_{1\rightarrow 0})$ for an internal gate, located inside a circuit. Initially, $\Pr(x_{i,0\rightarrow 1})$ and $\Pr(x_{i,1\rightarrow 0})$ are known for the primary inputs x_i of the circuit. In the core computational step of the algorithm, the $0 \rightarrow 1$ and $1 \rightarrow 0$ error components of input vectors at the inputs of a gate are combined to obtain the propagation error probabilities $P_{\text{p}}(g_{0\rightarrow 1})$ and $P_{\text{p}}(g_{1\rightarrow 0})$. These probabilities are then combined with the local gate failure probability to obtain $\Pr(g_{0\rightarrow 1})$ and $\Pr(g_{1\rightarrow 0})$ at the output of the gate.

The computation of propagation error probabilities $P_{\text{p}}(g_{0\rightarrow 1})$ and $P_{\text{p}}(g_{1\rightarrow 0})$ is described in the following. The single-pass reliability analysis is performed by recursively applying the core computational step of the algorithm to the gates in a topological order. At the end of the single pass, $\Pr(y_{0\rightarrow 1})$ and $\Pr(y_{1\rightarrow 0})$ are obtained for the output y of the circuit. The time complexity of the algorithm is $O(n)$, where n is the number of gates in the circuit. Note that the single-pass reliability analysis gives the exact values of probability of error at the output, in the absence of reconvergent fanout.

Expression for propagation error probabilities: The $0 \rightarrow 1$ and $1 \rightarrow 0$ input error probabilities at g are expressed as

$$\begin{aligned} P_{\text{p}}(g_{0\rightarrow 1}) &= \Pr(g_{0\rightarrow 1}|g \text{ is fault free}) \\ P_{\text{p}}(g_{1\rightarrow 0}) &= \Pr(g_{1\rightarrow 0}|g \text{ is fault free}) \end{aligned}. \tag{7.19}$$

These are probabilities that the output g would be erroneous (whereas the gate does not fail) for *at least* one input vector, i.e., probabilities for *combined* input error vectors. Expressions for the propagation error probabilities and its components, for a 2-input NAND gate with inputs i and j (gate labeled 3 in Fig. 7.6), are given in Table 7.2. P_{00}, P_{01}, P_{10}, and P_{11} represent the propagation error probability components for input vectors 00, 01, 10, and 11, respectively. $P_{00/01}$, $P_{00/10}$, $P_{01/10}$, and $P_{00/01/10}$ represent the joint propagation error probability components for 00/01, 00/10, 01/10, and 00/01/11 input vectors, respectively.

The calculation of $P_{\text{p}}(g_{1\rightarrow 0})$ to propagate the $1 \rightarrow 0$ error component using the entries in the upper part of Table 7.2 is described in the following; the propagation of the $0 \rightarrow 1$ input error component is similar, using the entries in the lower part of Table 7.2. Since the fault-free output value for $1 \rightarrow 0$ error is logic-1, there are seven rows in the upper table, one for each input vector and one for each possible combination of input vectors, for which the output of the fault-free NAND gate

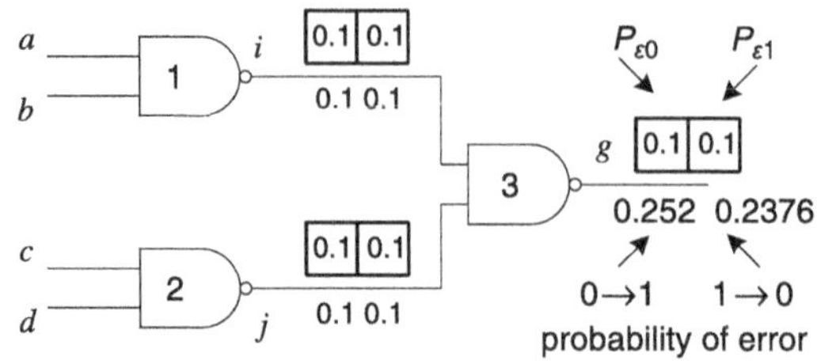

Fig. 7.6 Small circuit example realized with 2-input NAND gates used as a logic unit

Table 7.2 Expressions of input error components for 2-input NAND gate

Input vector	$1 \rightarrow 0$ input error component
00	$P_{00} = \Pr(i_{0\rightarrow 1})\Pr(j_{0\rightarrow 1})$
01	$P_{01} = \Pr(i_{0\rightarrow 1})(1 - \Pr(j_{1\rightarrow 0}))$
10	$P_{10} = \Pr(j_{0\rightarrow 1})(1 - \Pr(i_{1\rightarrow 0}))$
00/01	$P_{00/01} = \Pr(i_{0\rightarrow 1})\Pr(j_{0\rightarrow 1})(1 - \Pr(j_{1\rightarrow 0}))$
00/10	$P_{00/10} = \Pr(i_{0\rightarrow 1})\Pr(j_{0\rightarrow 1})(1 - \Pr(i_{1\rightarrow 0}))$
01/10	$P_{00/10} = \Pr(i_{0\rightarrow 1})\Pr(j_{0\rightarrow 1})(1 - \Pr(i_{1\rightarrow 0}))(1 - \Pr(j_{1\rightarrow 0}))$
00/01/10	$P_{00/10/10} = \Pr(i_{0\rightarrow 1})\Pr(j_{0\rightarrow 1})(1 - \Pr(i_{1\rightarrow 0}))(1 - \Pr(j_{1\rightarrow 0}))$
Total	$P_{\mathrm{p}}(g_{1\rightarrow 0}) = P_{00} + P_{01} + P_{10} - P_{00/01} - P_{00/10}$
Input vector	**$0 \rightarrow 1$ input error component**
11	$P_{11} = \Pr(i_{1\rightarrow 0}) + \Pr(j_{1\rightarrow 0}) - \Pr(i_{1\rightarrow 0})\Pr(j_{1\rightarrow 0})$
Total	$P_{\mathrm{p}}(g_{0\rightarrow 1}) = P_{11}$

is at logic-1. The first column in the table consists of the input vector or the joint group of input vectors under consideration. The input vector has been ordered as ij. The second column is the probability of a $1 \rightarrow 0$ error at g, only caused by errors at its inputs (when g itself is fault free). The entries in the second column are computed using $\Pr(i_{0\rightarrow 1})$, $\Pr(i_{1\rightarrow 0})$, $\Pr(j_{0\rightarrow 1})$, and $\Pr(j_{1\rightarrow 0})$, as illustrated below with examples.

Consider the input 10, whose error-free output is at logic-1. For g to be erroneous only due to errors at the inputs, j has to fail and i has to be error free so that the input to the gate is 11 instead of 10. Thus, the probability of a $1 \rightarrow 0$ error at g due to this input vector is $\Pr(j_{0\rightarrow 1})(1 - \Pr(i_{1\rightarrow 0}))$. Similar entries for the inputs 00 and 01 are derived.

Consider now the group consisting of input vectors 00 and 01. The joint probability of error for those two input vectors at g, i.e., the probability that g will be faulty due to both input vector failures, is the unknown. For g to be erroneous only due to errors at the inputs, i has to fail in both vectors; j has to fail if its default value is at logic-0 and has to be error free if its default value is at logic-1. Thus, the joint probability of a $1 \rightarrow 0$ error at g due to input vectors 00 and 01 is $\Pr(i_{0\rightarrow 1})\Pr(j_{0\rightarrow 1})(1 - \Pr(j_{1\rightarrow 0}))$. Similar entries for the groups of inputs vectors 00/10, 01/10, and 00/01/10 are derived.

The combined propagation error probability for $1 \rightarrow 0$, $P_{\mathrm{p}}(g_{1\rightarrow 0})$, is equal to the probability that *at least* one input vector is erroneous and is given in Table 7.2 as $P_{\mathrm{p}}(g_{1\rightarrow 0}) = P_{00} + P_{01} + P_{10} - P_{00/01} - P_{00/10} - P_{01/10} + P_{00/01/10}$. This expression is the *general addition rule* describing the probability that *any* of the multiple events (input vector failures in our case) occurs [246]:

$$P\left(\bigcup_{i=1}^{n} A_i\right) = \sum_{i=1}^{n} P(A_i) - \sum_{\substack{i,j=1\\ i<j}}^{n} P(A_i A_j) + \sum_{\substack{i,j,k=1\\ i<j<k}}^{n} P(A_i A_j A_k) + \cdots + (-1)^{n+1} P(A_1 A_2 \cdots A_n), \tag{7.20}$$

where $A_1, A_2, \ldots, A_n$ are non-mutually exclusive events.

Expressions for $\Pr(g_{0\to 1})$ *and* $\Pr(g_{1\to 0})$: If $P_{\epsilon 0}$ and $P_{\epsilon 1}$ are $0 \to 1$ and $1 \to 0$ probabilities of failure of the local gate output g, respectively, we get

$$\begin{aligned} \Pr(g_{0\to 1}) &= (1 - P_{\mathrm{p}}(g_{0\to 1})) \cdot P_{\epsilon 0} + P_{\mathrm{p}}(g_{0\to 1}) \cdot (1 - P_{\epsilon 1}), \\ \Pr(g_{1\to 0}) &= (1 - P_{\mathrm{p}}(g_{1\to 0})) \cdot P_{\epsilon 1} + P_{\mathrm{p}}(g_{1\to 0}) \cdot (1 - P_{\epsilon 0}). \end{aligned} \tag{7.21}$$

Note that the two terms $\Pr(j_{0\to 1})$ and $(1 - \Pr(i_{1\to 0}))$ are multiplied in the computation of the entries in the second column of Table 7.2. This implies that the events of i being correct and j failing are assumed independent. This assumption is valid if the gate is not a site of reconvergence of fanout. Since reconvergence causes the two events to be correlated, it is handled separately, further in this section. Moreover, the second column of Table 7.2 contains terms $\Pr(i_{0\to 1})(1 - \Pr(i_{1\to 0}))$ and $\Pr(j_{0\to 1})(1 - \Pr(j_{1\to 0}))$. Since $0 \to 1$ and $1 \to 0$ events are not independent, these factors should be replaced with $\Pr(i_{0\to 1}) - \Pr(i_{\mathrm{joint}})$ and $\Pr(j_{0\to 1}) - \Pr(j_{\mathrm{joint}})$ respectively, where $\Pr(i_{\mathrm{joint}})$ and $\Pr(j_{\mathrm{joint}})$ are the joint probabilities for $0 \to 1$ and $1 \to 0$ event on i and j, respectively. The total circuit's joint probability for $0 \to 1$ and $1 \to 0$ event is calculated in Section 7.2.1.2.

Although the computation has been illustrated for a NAND gate, the computation for a NOR gate is symmetric, i.e., there are seven rows for the probability of $0 \to 1$ and a single row for the probability of $1 \to 0$. Inverters, ANDs, ORs, and XORs are all handled in a similar manner and the tables have been excluded for brevity. Single-pass reliability analysis is illustrated for the circuit shown in Fig. 7.6. The gate failure probabilities ($P_{\epsilon 0}$ and $P_{\epsilon 1}$) and probabilities of $0 \to 1$ and $1 \to 0$ error are indicated for each gate. The gates are numbered in the order in which they are processed.

7.2.1.2 Joint Probability for Logic-0 and Logic-1 Errors

The single-pass reliability algorithm can be, furthermore, applied for joint probability calculation. The core calculation step is the same as previously explained. The difference lies in the propagation error probability calculation. The joint propagation error probability for $0 \rightarrow 1$ and $1 \rightarrow 0$ event ($P_{\mathrm{p}}(g_{\mathrm{joint}})$) is defined as

$$P_{\mathrm{p}}(g_{\mathrm{joint}}) = \Pr(g_{\mathrm{joint}} |\ g \text{ is free of simultaneous faults}), \tag{7.22}$$

where $\Pr(g_{\mathrm{joint}})$ is the probability of simultaneous failure for logic-0 and logic-1 ($0 \rightarrow 1$ and $1 \rightarrow 0$ event) at output of g. $P_{\mathrm{p}}(g_{\mathrm{joint}})$ is the probability that the output g would be erroneous (when gate does not fail) for *at least* one pair of input vectors, simultaneously, where one vector corresponds to default logic-0 at the output and another corresponds to default logic-1.

$P_{\epsilon 01}$ values of individual gates are acquired from MC simulations as $\Pr(y_{0\rightarrow 1} \text{and} y_{0\rightarrow 1})$, where y is the output of the individual gate when its inputs are fault free. The expressions of the components of propagation error probabilities, for a 2-input NAND gate with inputs i and j, are given in Table 7.3. The calculation of $P_{\mathrm{p}}(g_{\mathrm{joint}})$ is illustrated in an example.

Consider the group of input vectors 00, 01, and 11 (00/01 is a group that corresponds to default logic-0 at g and 11 corresponds to default logic-1 at g). The joint probability of error for those three input vectors at g is the unknown. For g to be erroneous only due to errors at the inputs, the following conditions need to be fulfilled: (i) $0 \rightarrow 1$ event has to occur for i for two input vectors and $1 \rightarrow 0$ event has to happen for one; this corresponds to the joint probability of failure – $\Pr(i_{\mathrm{joint}})$; (ii) j has to fail if its default value is at logic-0 and has to be error free if its default value is at logic-1, which corresponds to $\Pr(j_{0\rightarrow 1}) - \Pr(j_{\mathrm{joint}})$. Thus, the joint probability for all three input vectors at g is $\Pr(i_{\mathrm{joint}})(\Pr(j_{0\rightarrow 1}) - \Pr(j_{\mathrm{joint}}))$. Similar entries for the groups of inputs vectors 00/11, 01/11, 10/11, 00/10/11, 01/10/11, and 00/01/10/11 are derived. Notice that joint probabilities of failure for input vector groups 01/10/11 and 00/01/10/11 are equal to zero, since these probabilities depend on mutually exclusive events. $P_{\mathrm{p}}(g_{\mathrm{joint}})$ (given in Table 7.3) and $\Pr(g_{\mathrm{joint}})$ are calculated similarly as in Section 7.2.1.1.

$$\Pr(g_{\mathrm{joint}}) = (1 - P_{\mathrm{p}}(g_{\mathrm{joint}})) \cdot P_{\epsilon 01} + P_{\mathrm{p}}(g_{\mathrm{joint}}) \cdot (1 - P_{\epsilon 01}) \tag{7.23}$$

Table 7.3 Expressions for joint input error components for 2-input NAND gate

Input vector	Joint $0 \rightarrow 1$ and $1 \rightarrow 0$ input error component
00/11	$P_{00/11} = \Pr(i_{\mathrm{joint}})\Pr(j_{0\rightarrow 1}) + \Pr(i_{\mathrm{joint}})\Pr(i_{0\rightarrow 1}) - \Pr(i_{\mathrm{joint}})\Pr(i_{\mathrm{joint}})$
01/11	$P_{01/11} = \Pr(i_{\mathrm{joint}})(1 - \Pr(j_{1\rightarrow 0}))$
10/11	$P_{10/11} = \Pr(j_{\mathrm{joint}})(1 - \Pr(i_{1\rightarrow 0}))$
00/01/11	$P_{00/01/11} = \Pr(i_{\mathrm{joint}})(\Pr(j_{0\rightarrow 1}) - \Pr(j_{\mathrm{joint}}))$
00/10/11	$P_{00/10/11} = \Pr(j_{\mathrm{joint}})(\Pr(i_{0\rightarrow 1}) - \Pr(i_{\mathrm{joint}}))$
01/10/11	$P_{01/10/11} = 0$
00/01/10/11	$P_{00/01/10/11} = 0$
Total	$P_{\mathrm{p}}(g_{\mathrm{joint}}) = P_{00/11} + P_{01/11} + P_{10/11} - P_{00/01/11} - P_{00/10/11}$

7.2.1.3 Handling Reconvergent Fanout

Since the proposed algorithm is based on signal probabilities, its main drawback relates to signal correlations, which invalidate the straightforward computing of joint probabilities. The single-pass reliability algorithm computes the exact value of the probability of failure of circuits with no reconvergent fanouts. For practical circuits, computing the exact probabilities of a signal may be intractable, according to the number of reconvergent fanouts. The signal probabilities' problem is in the class of #P-complete (sharp P-complete) ones [215], possibly harder than NP-complete problems. Figure 7.7a shows an example of a circuit (ISCAS-C17 [249]) with two reconvergent fanouts and the exact probability of failure of the output z_2 for a NAND gate probability of failure $P_{\epsilon 0} = P_{\epsilon 1} = 0.1$. The signal e is also considered a reconvergent fanout since the circuit reliability computation involves the joint probabilities of the output signals. Figure 7.7b shows the equivalent circuit that is effectively computed and the calculated probability of failure of the output z_2, when these reconvergent fanout signals are not taken into account. When reconvergence is not taken into account, the obtained value for the probability of failure is *higher* compared to the exact result acquired through MC simulations.

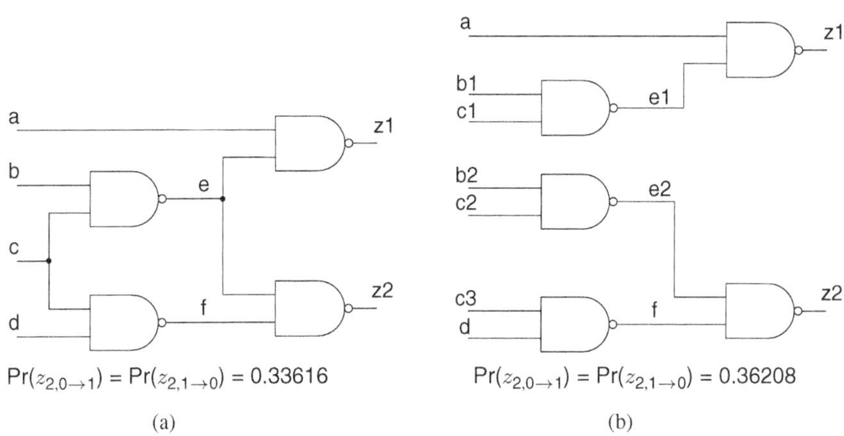

Fig. 7.7 (**a**) A circuit with a reconvergent fanout; (**b**) an equivalent circuit that is effectively computed when this reconvergence is not taken into account

The presence of reconvergent fanout renders the single-pass reliability analysis approximate because the events of $0 \to 1$ or $1 \to 0$ error for the inputs of a gate may not be independent at the point of reconvergence. Handling reconvergent fanout has been the subject of extensive research in signal probability computation. In this section, the theory of correlation coefficients used in signal probability computation [215] is extended to make the single-pass reliability analysis more accurate in the presence of reconvergent fanout. This approach relies on the propagation of the correlation coefficients for a pair of wires from the source of fanout to the point of reconvergence. Note that the word "wire" has been used as opposed to "node" because for a gate with fanout > 1, each fanout is treated as a separate wire, but they constitute the same node.

We define the correlation coefficient for events on a pair of wires as the joint probability of the events divided by the product of their marginal probabilities. For signal probability computation, an event on a wire is defined as the value of the wire being at logic-1. Thus, for a pair of wires, a single correlation coefficient is sufficient to compute the joint probability of a logic-1 on both wires.

In this analysis, an event is defined as a $0 \rightarrow 1$ or $1 \rightarrow 0$ error on a wire. Hence, instead of a single correlation coefficient, there are four correlation coefficients for a pair of wires, one for every combination of events on the pair of wires. Let v and w represent two wires. Four correlation coefficients for this pair are denoted by C_{vw}, $C_{\tilde{v}w}$, $C_{v\tilde{w}}$, and $C_{\tilde{v}\tilde{w}}$, where v, w, $\tilde{v}$, and $\tilde{w}$ refer to the event of a $0 \rightarrow 1$, $0 \rightarrow 1$, $1 \rightarrow 0$, and $1 \rightarrow 0$ error at v and w, respectively.

The correlation coefficients must be considered at the gates whose inputs are the site of reconvergence of fanout. At such gates, the events of a $0 \rightarrow 1$ or $1 \rightarrow 0$ error at the inputs are not independent. Thus, the entries in the second column of Table 7.2 are weighted by the appropriate correlation coefficient, e.g., $\Pr(i_{0\rightarrow 1})(1-\Pr(j_{1\rightarrow 0}))$ becomes $\Pr(i_{0\rightarrow 1})(1-\Pr(j_{1\rightarrow 0})C_{i\bar{j}})$.

Correlation coefficient computation: The correlation coefficient of a pair of wires can be calculated by computing the correlation coefficients of the wires in the fanout source that cause the correlation in the first step and then propagating these correlation coefficients along the appropriate paths leading to the pair of wires. Note that all four correlation coefficients for two independent wires are equal to 1. The computation of correlation coefficients for the fanout source and the propagation of correlation coefficients at a 2-input NAND gate are described in the following.

- *Computation at fanout source node*: The fanout source node i is shown in Fig. 7.8a. Following the definition of the correlation coefficient, the correlation coefficient for the pair of wires $\{l, m\}$ is computed as follows:

$$\begin{aligned} \Pr(l_{0\rightarrow 1}) = \Pr(l_{0\rightarrow 1}, m_{0\rightarrow 1}) &= \Pr(l_{0\rightarrow 1})\Pr(m_{0\rightarrow 1})C_{lm}, \\ C_{lm} &= \frac{1}{\Pr(m_{0\rightarrow 1})}. \end{aligned} \tag{7.24}$$

$C_{\tilde{l}\tilde{m}}$ can be computed in a similar manner. $C_{\tilde{l}m}$ and $C_{l\tilde{m}}$ are zero because it is not possible to have a $0 \rightarrow 1$ error on m and a $1 \rightarrow 0$ error on l, or vice versa.

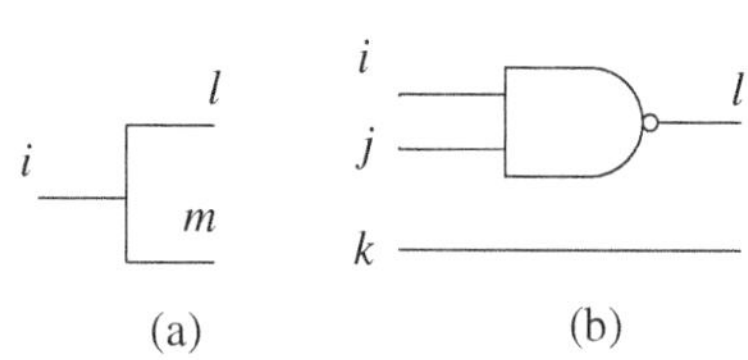

Fig. 7.8 Computation/propagation of correlation coefficient

- *Propagation at NAND gate*: The propagation of correlation coefficients is illustrated for the NAND gate in Fig. 7.8b. Let i, j, k be three wires whose pairwise correlation coefficients are known. The computation of the correlation coefficients for the pair $\{l, k\}$ involves the propagation of the correlation coefficients through the NAND gate, using the correlation coefficients for pairs (i, k) and (j, k). Following the definition of the correlation coefficient and the definition of the conditional probability [246]

$$C_{\tilde{l}k} = \frac{\Pr(l_{1\to 0}, k_{0\to 1})}{\Pr(l_{1\to 0})\Pr(k_{1\to 0})} = \frac{\Pr(l_{1\to 0}|k_{0\to 1})}{\Pr(l_{1\to 0})}. \tag{7.25}$$

The expression of $\Pr(l_{1\to 0}|k_{0\to 1})$ in terms of the correlation coefficients for pairs of inputs (i, k) and (j, k) is given as

$$\begin{aligned}
\Pr(l_{1\to 0}|k_{0\to 1}) &= (1 - P_p(l_{1\to 0}|k_{0\to 1})) \cdot P_{\epsilon 0} + P_p(l_{1\to 0}|k_{0\to 1}) \cdot P_{\epsilon 1} \\
&\text{and} \\
P_{\mathrm{p}}(l_{1\to 0}|k_{0\to 1}) &= \Pr(i_{0\to 1}|k_{0\to 1})\Pr(j_{0\to 1}|k_{0\to 1})C_{ij} \\
&\quad + \Pr(i_{0\to 1}|k_{0\to 1})(1 - \Pr(j_{1\to 0}|k_{0\to 1}))C_{i\tilde{j}} \\
&\quad + \Pr(j_{0\to 1}|k_{0\to 1})(1 - \Pr(i_{1\to 0}|k_{0\to 1}))C_{\tilde{i}j} \\
&\quad + \Pr(i_{0\to 1}|k_{0\to 1})\Pr(j_{0\to 1}|k_{0\to 1})(1 - \Pr(j_{1\to 0}|k_{0\to 1}))C_{ij} \\
&\quad + \Pr(i_{0\to 1}|k_{0\to 1})\Pr(j_{0\to 1}|k_{0\to 1})(1 - \Pr(i_{1\to 0}|k_{0\to 1}))C_{ij} \\
&= \Pr(i_{0\to 1})C_{ik}\Pr(j_{0\to 1})C_{jk}C_{ij} \\
&\quad + \Pr(i_{0\to 1})C_{ik}(1 - \Pr(j_{1\to 0}))C_{\tilde{j}k}C_{i\tilde{j}} \\
&\quad + \Pr(j_{0\to 1})C_{jk}(1 - \Pr(i_{1\to 0}))C_{\tilde{i}k}C_{\tilde{i}j} \\
&\quad + \Pr(i_{0\to 1})C_{ik}\Pr(j_{0\to 1})C_{jk}(1 - \Pr(j_{1\to 0}))C_{\tilde{j}k}C_{ij} \\
&\quad + \Pr(i_{0\to 1})C_{ik}\Pr(j_{0\to 1})C_{jk}(1 - \Pr(i_{1\to 0}))C_{\tilde{i}k}C_{ij}.
\end{aligned} \tag{7.26}$$

The terms in the expression of $P_{\mathrm{p}}(l_{1\to 0}|k_{0\to 1})$ are similar to the terms in the second column of the upper part of Table 7.2. The difference is that the probability of $0 \to 1$ and $1 \to 0$ errors has been multiplied by appropriate correlation coefficients. However, the terms that have more than two probability factors (fourth to seventh component from Table 7.2) are not fully accurate because correlation factors are defined only pairwise. Still, these terms are small compared to the first three components that are fully accurate. The expression of C_{lk} is derived in a similar manner, using the lower part of Table 7.2. Expressions of $C_{l\tilde{k}}$ and $C_{\tilde{l}\tilde{k}}$ are derived by replacing k with $\tilde{k}$ in the expressions of C_{lk} and $C_{\tilde{l}k}$, respectively.

In comparison with the original single-pass reliability analysis tool presented in [212], the modified version has the same level of accuracy and slightly improved speed. The speed improvement is due to the omission of weight vector calculation. The modified single-pass reliability evaluation tool presents a fast, accurate, and scalable solution that provides the $0 \rightarrow 1$ and $1 \rightarrow 0$ probability of failure for the output of an arbitrary gate in the circuit. This information is crucial for the output PDF modeling procedure that is presented in the following section.

7.2.2 Output PDF Modeling

The procedure for modeling PDFs of the worst-case logic-0 and logic-1 at the output of an arbitrary gate g in the circuit is derived. The same example circuit is used, as in the previous section (Fig. 7.6). If the desired PDFs are marked with $h_{g,\max 0}$ and $h_{g,\min 1}$, the probability of failure acquired using the single-pass reliability tool is given according to (7.14) as

$$\mathrm{Pr}(g_{0\rightarrow 1}) = \int_{0.5}^{1} h_{g,\max 0}(x)dx \quad \text{and} \quad \mathrm{Pr}(g_{1\rightarrow 0}) = \int_{0}^{0.5} h_{g,\min 1}(x)dx, \tag{7.27}$$

since all output values for logic-0 (logic-1) that are higher (lower) than a threshold of 0.5 are assumed as erroneous.

Output PDFs are modeled using the two values for the probability of the gate failure from (7.27) acquired with the single-pass reliability tool and PDFs of individual gates that are close to output in a topological sense. $0 \rightarrow a$ marks an event where the output of the gate is at any level that is different from logic-0 and logic-1 $(0 < a < 1)$ when its fault-free value is logic-0. Accordingly, $1 \rightarrow a$ marks an event where the output of the gate is at any level different from logic-0 and logic-1 $(0 < a < 1)$ when its fault-free value is logic-1. We differentiate four distinctive cases of the propagation error probability at g

$$\begin{aligned} P_{\mathrm{p}}(g_{0\rightarrow 1}) &= \mathrm{Pr}(g_{0\rightarrow 1} | \, g \text{ is fault free}) \\ P_{\mathrm{p}}(g_{0\rightarrow a}) &= \mathrm{Pr}(g_{0\rightarrow a} | \, g \text{ is fault free}) \\ P_{\mathrm{p}}(g_{1\rightarrow 0}) &= \mathrm{Pr}(g_{1\rightarrow 0} | \, g \text{ is fault free}) \\ P_{\mathrm{p}}(g_{1\rightarrow a}) &= \mathrm{Pr}(g_{1\rightarrow a} | \, g \text{ is fault free}) \end{aligned}. \tag{7.28}$$

If the PDF of the local gate output g for the worst-case logic-0 (logic-1) is marked as $h_{\epsilon,\max 0}$ ($h_{\epsilon,\min 1}$) and $h_{\mathrm{p},0\rightarrow a}$ ($h_{\mathrm{p},1\rightarrow a}$), the PDF at the output of g for $0 \rightarrow a$

$(1 \rightarrow a)$ propagation error $h_{g,\max 0}$ and $h_{g,\min 1}$ is then expressed as

$$\begin{aligned} h_{g,\max 0} &= (1 - P_{\rm p}(g_{0\rightarrow 1}) - P_{\rm p}(g_{0\rightarrow a})) \cdot h_{\epsilon,\max 0} \\ &\quad + P_{\rm p}(g_{0\rightarrow 1}) \cdot h_{\epsilon,\min 1} + P_{\rm p}(g_{0\rightarrow a}) \cdot h_{{\rm p},0\rightarrow a}, \\ h_{g,\min 1} &= (1 - P_{\rm p}(g_{1\rightarrow 0}) - P_{\rm p}(g_{1\rightarrow a})) \cdot h_{\epsilon,\min 1} \\ &\quad + P_{\rm p}(g_{1\rightarrow 0}) \cdot h_{\epsilon,\max 0} + P_{\rm p}(g_{1\rightarrow a}) \cdot h_{{\rm p},1\rightarrow a}. \end{aligned} \quad (7.29)$$

$h_{{\rm p},0\rightarrow a}$ and $h_{{\rm p},1\rightarrow a}$ consist of two components: one for a fault-free g and another for a faulty g:

$$\begin{aligned} h_{{\rm p},0\rightarrow a} &= (1 - P_{\epsilon 0}) \cdot h^{(1)}_{{\rm p},0\rightarrow a} + P_{\epsilon 0} \cdot h^{(2)}_{{\rm p},0\rightarrow a}, \\ h_{{\rm p},1\rightarrow a} &= (1 - P_{\epsilon 1}) \cdot h^{(1)}_{{\rm p},1\rightarrow a} + P_{\epsilon 1} \cdot h^{(2)}_{{\rm p},1\rightarrow a}, \end{aligned} \quad (7.30)$$

where $h^{(1)}_{{\rm p},0\rightarrow a}$ ($h^{(1)}_{{\rm p},1\rightarrow a}$) is the PDF at the output of g for the $0 \rightarrow a$ $(1 \rightarrow a)$ event when g is fault free and $h^{(2)}_{{\rm p},0\rightarrow a}$ ($h^{(2)}_{{\rm p},1\rightarrow a}$) is the PDF at the output of g for the $0 \rightarrow a$ $(1 \rightarrow a)$ event when g is faulty. After inserting (7.30) into (7.29)

$$\begin{aligned} h_{g,\max 0} &= (1 - P_{\rm p}(g_{0\rightarrow 1}) - P_{\rm p}(g_{0\rightarrow a})) \cdot h_{g,\max 0} + P_{\rm p}(g_{0\rightarrow 1}) \cdot h_{g,\min 1} \\ &\quad + P_{\rm p}(g_{0\rightarrow a}) \cdot (1 - P_{\epsilon 0}) \cdot h^{(1)}_{{\rm p},0\rightarrow a} + P_{\rm p}(g_{0\rightarrow a}) \cdot P_{\epsilon 0} \cdot h^{(2)}_{{\rm p},0\rightarrow a}, \\ h_{g,\min 1} &= (1 - P_{\rm p}(g_{1\rightarrow 0}) - P_{\rm p}(g_{1\rightarrow a})) \cdot h_{g,\min 1} + P_{\rm p}(g_{1\rightarrow 0}) \cdot h_{g,\max 0} \\ &\quad + P_{\rm p}(g_{1\rightarrow a}) \cdot (1 - P_{\epsilon 1}) \cdot h^{(1)}_{{\rm p},1\rightarrow a} + P_{\rm p}(g_{1\rightarrow a}) \cdot P_{\epsilon 1} \cdot h^{(2)}_{{\rm p},1\rightarrow a}, \end{aligned} \quad (7.31)$$

where $P_{\epsilon 0}$ and $P_{\epsilon 1}$ are probabilities of failure of the local gate output g for logic-0 and logic-1, respectively, as presented at the beginning of Section 7.2.1. Not all the elements of the sum in (7.31) have the same impact on $h_{g,\max 0}$ ($h_{g,\min 1}$). The impact is determined by probability factors that multiply PDFs on the right-hand side of (7.31), since the integral of each of these PDFs is equal to one. The probability factors ratio is given as

$$\begin{aligned} &1 - P_{\rm p}(g_{0\rightarrow 1}) - P_{\rm p}(g_{0\rightarrow a}) \gg P_{\rm p}(g_{0\rightarrow 1}) > \\ &\qquad > P_{\rm p}(g_{0\rightarrow a}) \cdot (1 - P_{\epsilon 0}) \gg P_{\rm p}(g_{0\rightarrow a}) \cdot P_{\epsilon 0}, \\ &1 - P_{\rm p}(g_{1\rightarrow 0}) - P_{\rm p}(g_{1\rightarrow a}) \gg P_{\rm p}(g_{1\rightarrow 0}) > \\ &\qquad > P_{\rm p}(g_{1\rightarrow a}) \cdot (1 - P_{\epsilon 1}) \gg P_{\rm p}(g_{1\rightarrow a}) \cdot P_{\epsilon 1}. \end{aligned} \quad (7.32)$$

The smallest propagation factor (the last element on the right-hand side of (7.32)) can be neglected. After introducing this approximation, (7.31) becomes

$$\begin{aligned} h_{g,\max 0} &\approx (1 - P_{\mathrm{p}}(g_{0\to 1}) - P_{\mathrm{p}}(g_{0\to a})) \cdot h_{\epsilon,\max 0} \\ &\quad + P_{\mathrm{p}}(g_{0\to 1}) \cdot h_{\epsilon,\min 1} + P_{\mathrm{p}}(g_{0\to a}) \cdot h^{(1)}_{\mathrm{p},0\to a}, \\ h_{g,\min 1} &\approx (1 - P_{\mathrm{p}}(g_{1\to 0}) - P_{\mathrm{p}}(g_{1\to a})) \cdot h_{\epsilon,\min 1} \\ &\quad + P_{\mathrm{p}}(g_{1\to 0}) \cdot h_{\epsilon,\max 0} + P_{\mathrm{p}}(g_{1\to a}) \cdot h^{(1)}_{\mathrm{p},1\to a}. \end{aligned} \tag{7.33}$$

The unknowns in (7.33) are $P_{\mathrm{p}}(g_{0\to a}) \cdot h^{(1)}_{\mathrm{p},0\to a}$ and $P_{\mathrm{p}}(g_{1\to a}) \cdot h^{(1)}_{\mathrm{p},1\to a}$. All other elements can be derived.

To determine the unknowns, the transformation of PDFs is observed at the input of g by the gate transfer function. The transfer function is determined for a fault-free library gate with default (output gate) load. Similar to the previous section, the 2-input NAND gate is considered without losing generality, and the circuit example is depicted in Fig. 7.6. A single-input transfer function for a typical 2-input NAND gate is depicted in Fig. 7.9a. In order for the $0 \to a$ $(1 \to a)$ event to occur, one input has to be in the region defined as $[\Delta_a, V_{\mathrm{DD}} - \Delta_a]$, with fault-free value at logic-1 (logic-0), and another input has to be at logic-1. If PDFs for the worst-case logic-0 and logic-1 at inputs i and j of g are marked as $h_{i,\max 0}$, $h_{i,\min 1}$, $h_{j,\max 0}$, and $h_{j,\min 1}$, respectively, and the gate g transfer function of the PDF in the region of interest ($[\Delta_a, V_{\mathrm{DD}} - \Delta_a]$) is f_{T}

$$\begin{aligned} P_{\mathrm{p}}(g_{0\to a}) \cdot h^{(1)}_{\mathrm{p},0\to a} &= f_{\mathrm{T}}(h_{i,\min 1})(1 - \Pr(j_{1\to 0})) + f_{\mathrm{T}}(h_{j,\min 1})(1 - \Pr(i_{1\to 0})), \\ P_{\mathrm{p}}(g_{1\to a}) \cdot h^{(1)}_{\mathrm{p},1\to a} &= f_{\mathrm{T}}(h_{i,\max 0})(1 - \Pr(j_{1\to 0})) + f_{\mathrm{T}}(h_{j,\max 0})(1 - \Pr(i_{1\to 0})). \end{aligned} \tag{7.34}$$

$h_{i,\max 0}$, $h_{i,\min 1}$, $h_{j,\max 0}$, and $h_{j,\min 1}$ also respect (7.33). f_{T} is computed numerically using the gate g transfer function. The PDF of the 2-input NAND gate in the region of interest ($[\Delta_a, V_{\mathrm{DD}} - \Delta_a]$) (depicted in green in Fig. 7.9b) h_a and its transformation through the transfer function $f_{\mathrm{T}}(h_a)$ are given in Fig. 7.9b, c, respectively. In the transformed PDF $f_{\mathrm{T}}(h_a)$, the probability values in the $[\Delta_a, V_{\mathrm{DD}} - \Delta_a]$ region are significantly smaller than the values outside this region. Taking this fact into consideration, $h_{i,\max 0}$ and $h_{i,\min 1}$ ($h_{j,\max 0}$ and $h_{j,\min 1}$) which also comply with (7.33) can be approximated by $h_{\epsilon,\max 0}$ and $h_{\epsilon,\min 1}$ in the region of interest. This approximation means that $P_{\mathrm{p}}(g_{0\to a}) \cdot h^{(1)}_{\mathrm{p},0\to a}$ and $P_{\mathrm{p}}(g_{1\to a}) \cdot h^{(1)}_{\mathrm{p},1\to a}$ *only depend on the PDFs of individual gates in the last layer of the fanin cone* (outputs i and j in Fig. 7.6) and that the propagating factors in these PDFs can be neglected. Since $P_{\mathrm{p}}(g_{0\to a})$ ($P_{\mathrm{p}}(g_{1\to a})$) is the smallest factor in (7.33), this approximation is justified. For the same reason, factors $\Pr(i_{1\to 0})$ and $\Pr(j_{1\to 0})$ are omitted from (7.34). Finally,

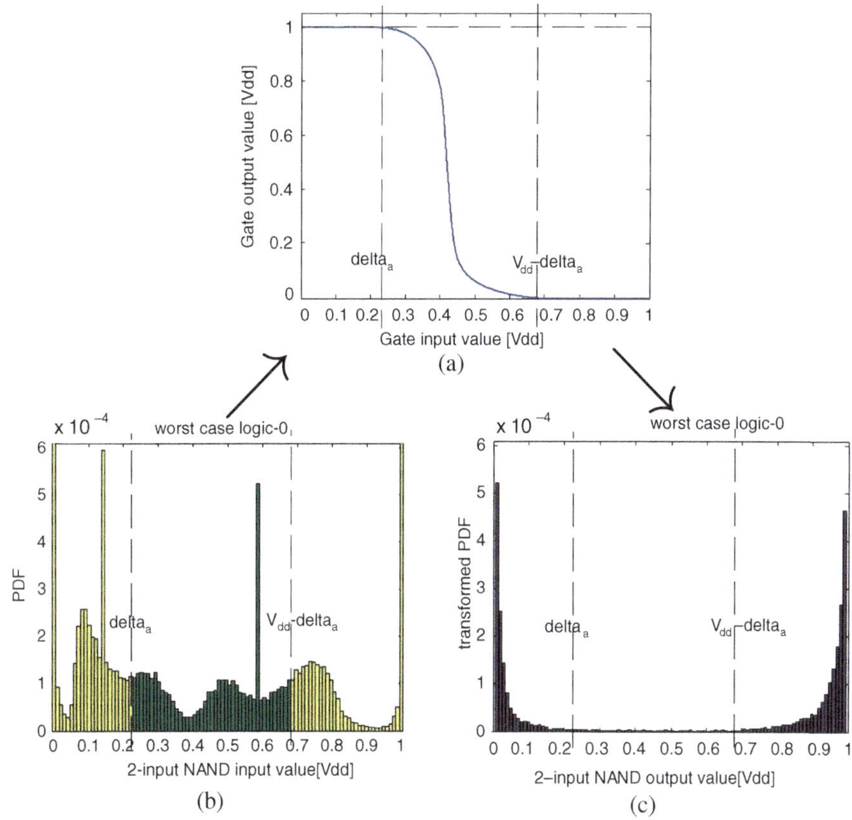

Fig. 7.9 2-input NAND. (**a**) gate transfer function; (**b**) PDF for the worst-case logic-0 in the $[\triangle_a, V_{\mathrm{DD}} - \triangle_a]$ region; (**c**) transformation of PDF from (**b**) through gate transfer function from (**a**)

$$
\begin{aligned}
P_{\mathrm{p}}(g_{0\to a}) \cdot h^{(1)}_{\mathrm{p},0\to a} &\approx 2 f_{\mathrm{T}}(h_{\epsilon,\min 1}), \\
P_{\mathrm{p}}(g_{1\to a}) \cdot h^{(1)}_{\mathrm{p},1\to a} &\approx 2 f_{\mathrm{T}}(h_{\epsilon,\max 0}).
\end{aligned}
\tag{7.35}
$$

From (7.35), $P_{\mathrm{p}}(g_{0\to a})$ and $P_{\mathrm{p}}(g_{1\to a})$ are expressed as

$$
\begin{aligned}
P_{\mathrm{p}}(g_{0\to a}) &\approx \int_0^1 2 f_{\mathrm{T}}(h_{\epsilon,\min 1}), \\
P_{\mathrm{p}}(g_{1\to a}) &\approx \int_0^1 2 f_{\mathrm{T}}(h_{\epsilon,\max 0}).
\end{aligned}
\tag{7.36}
$$

After substituting (7.33) into (7.27) and solving for $P_{\mathrm{p}}(g_{0\to 1})$ and $P_{\mathrm{p}}(g_{1\to 0})$, the remaining unknown factors from (7.33) are derived as

$$P_{\mathrm{p}}(g_{0\to 1}) = \frac{\Pr(g_{0\to 1}) - (1 - P_{\mathrm{p}}(g_{0\to a}))P_{\epsilon 0} - P_{\mathrm{p}}(g_{0\to a})\int_{0.5}^{1} h^{(1)}_{\mathrm{p},0\to a}(x)dx}{1 - P_{\epsilon 0} - P_{\epsilon 1}},$$
$$P_{\mathrm{p}}(g_{1\to 0}) = \frac{\Pr(g_{1\to 0}) - (1 - P_{\mathrm{p}}(g_{1\to a}))P_{\epsilon 1} - P_{\mathrm{p}}(g_{1\to a})\int_{0}^{0.5} h^{(1)}_{\mathrm{p},1\to a}(x)dx}{1 - P_{\epsilon 0} - P_{\epsilon 1}}. \tag{7.37}$$

A MC framework performing SPICE-level simulations is used to obtain data for the comparison and the necessary PDFs of the library gates. All possible combinations of single and double permanent faults are injected in each standard library gate, in order to generate PDFs for values of the probability of fault per transistor (p_{f}) in the range from 0.1 to 20%. A *4-bit full adder* is used as the main benchmark circuit. This is an area/delay minimized realization of an adder, synthesized in Synopsis using the reduced library set consisting of 2- and 3-input NAND and NOR gates and inverters. The benchmark circuit consists of 39 gates in total.

The modeled PDFs for the worst-case logic-0 and logic-1 at outputs of *4-bit full adder* (denoted as "modeled" in the figures) are compared with the equivalent PDFs acquired using MC tool (denoted as "simulated" in the figures). Single, double, and triple permanent faults have been injected into the MC framework. The number of MC iterations is selected as 3×320,000 in order to minimize the error due to sampling.

The values of modeled and simulated PDFs for the worst-case logic-0 are depicted in Fig. 7.10a, b. Likewise, the values of modeled and simulated PDFs for the worst-case logic-1 are depicted in Fig. 7.11a, b. The applied p_{f} to obtain data depicted in the figures is 2%. All values are depicted in 100 bins original histograms (without interpolation). The values of histograms for the output equal to zero and V_{DD} are excluded for better visualization, since they are few orders of magnitude larger than other values in the histogram. The difference between PDFs is not noticeable.

The intrinsic property of our method is that the runtime does not depend on the complexity of the circuit once the circuit output probability of failure is acquired. The average runtime of our tool for the benchmark circuit (for different values of p_{f}), including single-pass reliability analysis tool, is under 100 ms.

In order to compare modeled and simulated PDFs, Pearson's chi-square test [246] for histogram comparison has been performed. The chi-square statistics calculates the difference between simulated and modeled histograms as

$$X^2 = \sum_{i=1}^{r} \frac{(S_i - nM_i)^2}{nM_i}, \tag{7.38}$$

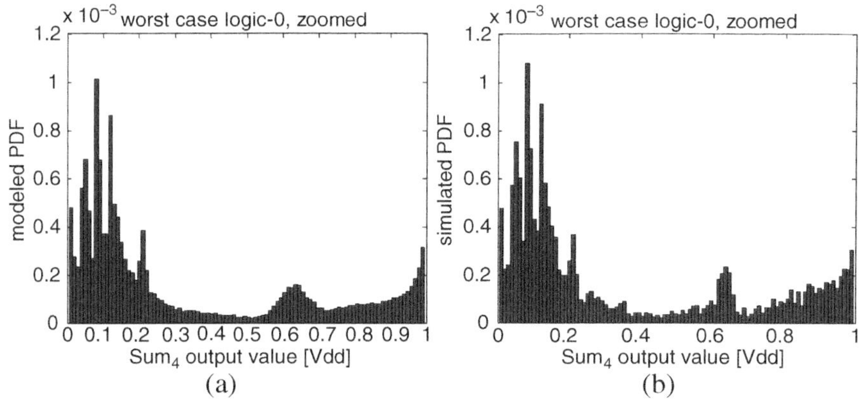

Fig. 7.10 *4-bit full-adder* worst-case logic-0 PDF (zoomed): (**a**) modeled; (**b**) simulated

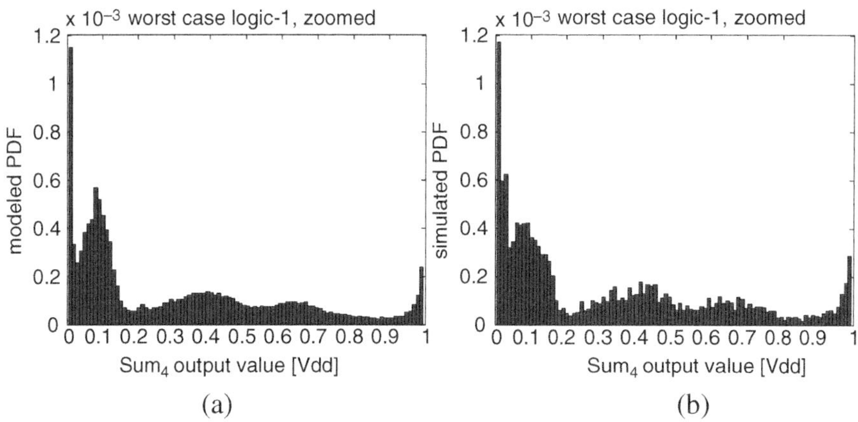

Fig. 7.11 *4-bit full-adder* worst-case logic-1 PDF (zoomed): (**a**) modeled; (**b**) simulated

where X^2 is the test statistic that asymptotically approaches a χ^2 distribution, S_i is the simulated histogram value (directly acquired from MC simulations), M_i is the modeled histogram value (normalized; the sum of its values is equal to one), n is the number of iterations (only the iterations when circuit output was faulty are included), and r is the number of histogram bins (100 in our case). The null hypothesis states that the compared distributions are identical. This hypothesis can be rejected with significance level α (the rejecting error of the correct null hypothesis) if and only if $X^2 > \varepsilon_{1-\alpha}$, where $\varepsilon_{1-\alpha}$ is $1 - \alpha$ quantile of the $\chi^2(r-1)$ distribution.

Table 7.4 shows X^2 values for PDFs of each output of the *4-bit full adder* (Sum_1 to Sum_4) for the worst-case logic-0 and logic-1 and for different values of p_f. The 1 and 5% quantile values for $\varepsilon_{1-\alpha}$ used for comparison are 134.642 and 123.225, respectively. Following the chi-square test results, the hypothesis that modeled and

Table 7.4 Chi-square test results: X^2 values for outputs of *4-bit full adder* for the worst-case logic-0 and logic-1 and for different values of p_f

		p_f=0.1%	p_f=1%	p_f=5%	p_f=20%
Output		X^2	X^2	X^2	X^2
Sum_1	logic-0	35.72	38.49	28.15	34.93
	logic-1	23.30	29.88	19.41	31.23
Sum_2	logic-0	68.13	62.56	61.98	72.54
	logic-1	51.86	60.06	57.33	50.43
Sum_3	logic-0	44.24	38.42	41.79	47.62
	logic-1	61.47	65.07	69.51	62.83
Sum_4	logic-0	55.29	54.57	51.84	59.90
	logic-1	76.43	82.74	74.86	81.03

simulated distributions are identical cannot be rejected for any significance level and for any of the compared PDFs. All X^2 values are smaller than the value of $\varepsilon_{0.95}$. X^2 value does not change noticeably for different p_f.

The novel statistical method for modeling an arbitrary circuit output PDFs is presented in this section. The method enables accurate modeling (confirmed with Pearson's chi-square test) of PDFs for the worst-case logic-0 and logic-1 of the unit output. The method is implemented in MATLAB scripts and is very fast; all calculations presented in Table 7.1 are performed in less than 100 ms. To the best of our knowledge, this is the first work that presents fast and accurate modeling of output PDFs of an arbitrary circuit. The method has been demonstrated in standard CMOS technology for permanent transistor defects modeled with detailed transistor fault models. However, there are no restriction on the application of the method to any fabrication technology, including future nanodevices. Considering the applied fault types, the method is not restricted to permanent defects and can be applied for transient faults in an equal manner.

7.3 Conclusions

The need for EDA tools that offer realistic reliability evaluation and modeling by employing a data collection without oversimplification of the models is becoming more prominent. The precise evaluation of the reliability of logic circuits has a significant importance not only because of the possibility to compare different fault-tolerant techniques, but also because the circuit design in highly defective and future nanotechnologies can be enhanced.

- A novel general method for fast and accurate statistical analysis of averaging fault-tolerant techniques has been presented in this chapter. The method consists of two important steps: (i) the advanced single-pass reliability evaluation that is used for modeling the PDF of an arbitrary gate and (ii) the statistical method for the analysis of fault-tolerant techniques that uses the acquired PDFs to provide reliability figures.

- The output PDFs of an arbitrary gate are only dependent on PDFs of gates that are located in the last two layers of the fanout cone and on output probability of failure. This enabled the development of the advanced single-pass reliability evaluation method.
- The novel advanced single-pass reliability evaluation method performs fast and accurate modeling of output PDFs of an arbitrary circuit. The accuracy of the method has been demonstrated in standard CMOS technology for permanent transistor defects modeled with detailed transistor fault models.
- The statistical method for the analysis of fault-tolerant techniques has also been presented and verified using data obtained by means of MC simulations. The output PDF of different decision gates used in conjunction with R redundant units is obtained using mathematical transformations of the unit output PDFs. Finally, the probability of error is derived from the output PDF of the decision gate.
- The importance of the method reflects in its enabling comparison and optimization of fault-tolerant techniques at higher level of abstraction and fulfilling prerequisites for system-level reliability evaluation and optimization that are presented in Chapter 8.

The accuracy of the PDF modeling method mainly depends on the accuracy of the modified single-pass reliability analysis tool. The accuracy of the tool depends on the algorithm which handles the reconverging fanouts. In the existing realization this algorithm implements pairwise correlation coefficients evaluation. Therefore, development of higher order correlation coefficients and inclusion in the tool without larger speed penalty could significantly improve the accuracy.

Chapter 8
Design Methodology: Reliability Evaluation and Optimization

Nowadays, standard design flows for digital logic design rely on optimization of important parameters such as speed, area, and power. However, even though reliability has been demonstrated as an important parameter that needs to be addressed in the design process, its optimization has not yet found the way into state-of-the-art design approaches.

In order to provide the end-user digital IC designer with a full reliability-aware design flow, an adaptation of the standard design flow which is applied nowadays is proposed according to Fig. 8.1. The proposed design flow intends to bridge the gap between the design methodology for nanodevices and existing design methodologies that are dealing with "micro"-scale CMOS devices by including automated reliability evaluation and optimization steps into a standard design flow. From an end-user's point of view, the approach should not differ much from today's design flows, so it is justified to consider that a new methodology should represent an upgrade to the existing one. The reliability evaluation and optimization steps should remain transparent for the end-user, who should have control over the process through additional reliability constraints.

Following the well-established hierarchical approach in order to deal with the complexity of the system (chip), a two-level reliability evaluation and optimization is performed, i.e., at (i) local level and (ii) system level. The accuracy of evaluation and the efficiency of optimization are significantly higher at a local level compared to a system level. However, due to an increased computational complexity, the approach applied at system level takes into consideration specific approximations that make evaluation and optimization achievable within an acceptable time constraint. Even though evaluation and optimization are performed off-line, runtime is still an important constraint. The distinction between local level and system level is also dependent on the evaluation and optimization procedure which is basically a two-step process. The local level is assumed as a single unit to which a basic replication is applied, together with a particular type of a decision gate forming a reliable block. At the system level, these reliable blocks are combined to build the whole system (chip) applying fault-tolerant techniques. The reliability of the chip is evaluated and reliable blocks are optimized with respect to the size and the redundancy factor. The local and system level examples are illustrated in Fig. 8.2,

M. Stanisavljević et al., *Reliability of Nanoscale Circuits and Systems*,
DOI 10.1007/978-1-4419-6217-1_8, © Springer Science+Business Media, LLC 2011

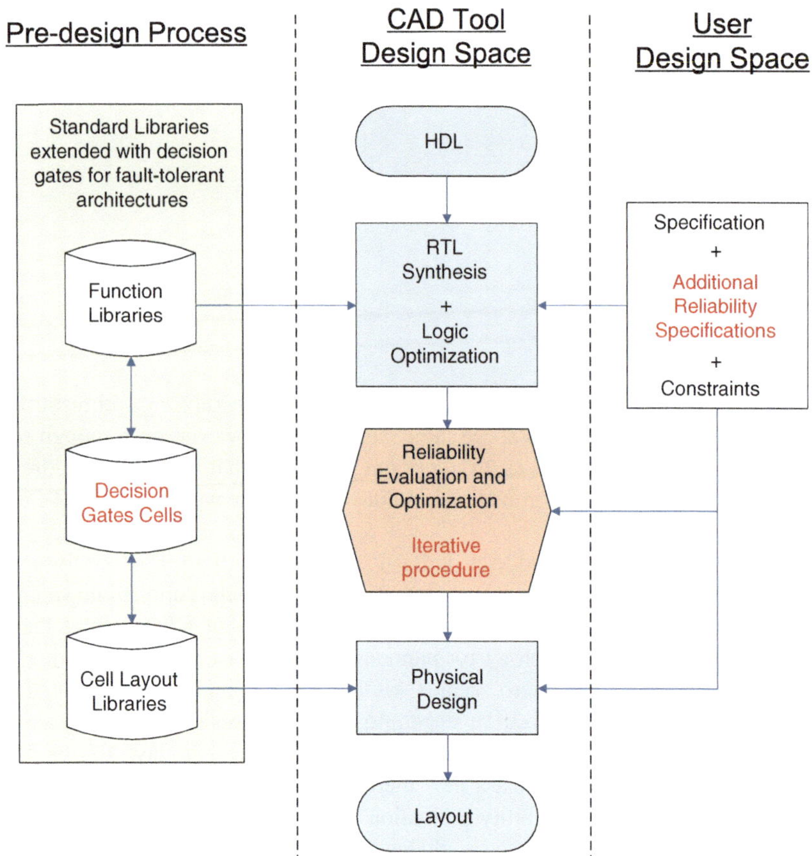

Fig. 8.1 Fault-tolerant design methodology flow as an upgrade of a standard design flow

with two examples at local level (gate level and extended gate level) depending on the optimal partition size.

Reliability evaluation and optimization as a part of the fault-tolerant design methodology (Fig. 8.1) is an iterative procedure based on the improvement of the accuracy of the estimation of the unit's average reliability, in each step. In the first step, the evaluation of the system reliability is performed taking into consideration the initial estimation of the unit's average reliability performed according to the unit's size and logic depth. Then the optimal partition size and redundancy factor determined. The actual system is optimally partitioned following design constraints. During partitioning itself, the average reliability of partitions is also optimized. After partitioning, a sufficient sample of partitions is chosen to be more accurately analyzed using the advanced single-pass reliability evaluation tool. The acquired reliability values are used to re-optimize the design and recalculate the optimal partition size and redundancy factor. If the difference from the initial partition size and redundancy factor is small enough, the procedure is over, otherwise a new iteration

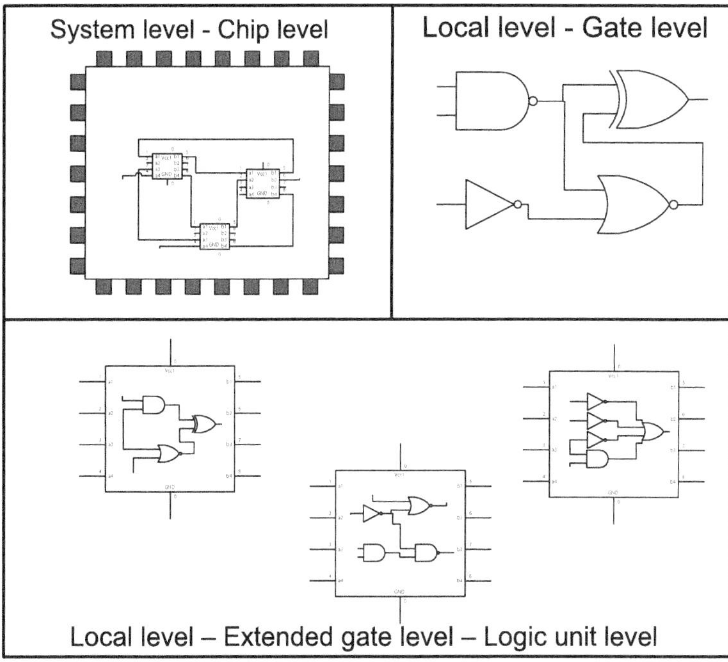

Fig. 8.2 System- and local-level illustration

is performed. The procedure has been illustrated in Fig. 8.3. In this chapter, each step of the reliability evaluation and optimization (local-level reliability evaluation, reliability-optimal partitioning, and system-level reliability evaluation and optimization) is addressed in detail.

8.1 Local-Level Reliability Evaluation

The accurate estimation and evaluation of the local-level reliability is crucial for subsequent system-level reliability evaluation and optimization. The probability of failure of each output of a unit mainly depends on the logic depth of its critical paths, as demonstrated later in this section. On the other hand, in terms of reliability, the improvement compared to majority voter-based techniques is significantly reduced with increased logic depth. These two dependencies are investigated in detail using a large number of sample circuits that have been evaluated for reliability and effectiveness of the averaging techniques.

The sample circuits used in the following analysis are obtained by partitioning a large design (12-bit look-up table) into various circuits of different logic depths, ranging from 2 to 15. The partitioning is performed using a customized partitioner based on hMetis [250] that has a logic depth minimization goal. A 12-bit look-up table that performs a bijective function, mapping each 12-bit input into one

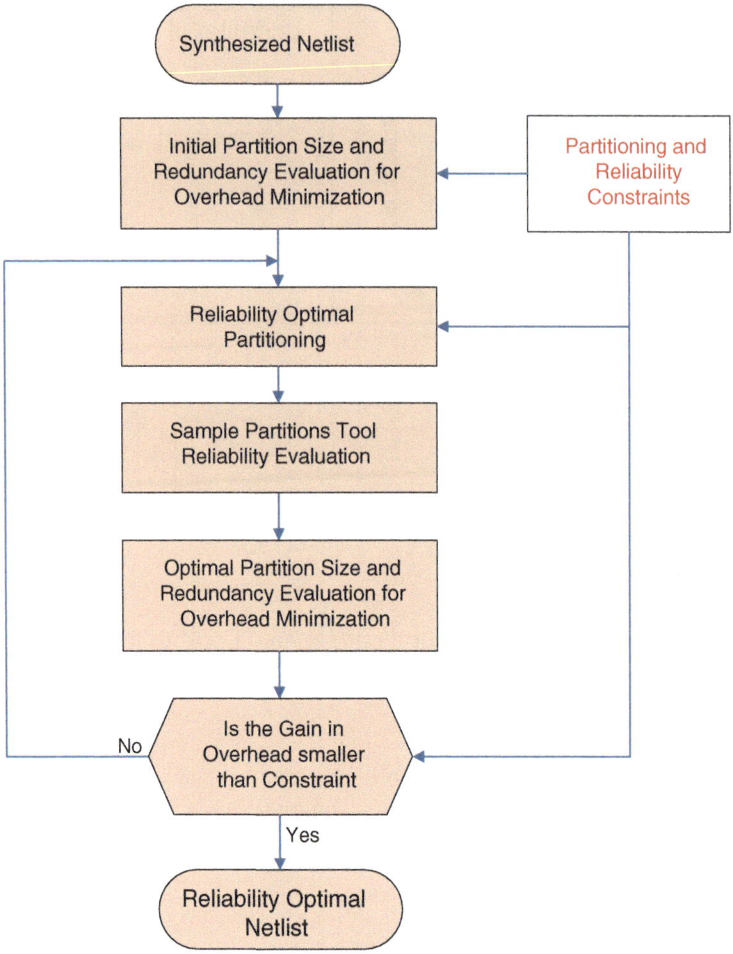

Fig. 8.3 Reliability evaluation and optimization procedure

12-bit output, has been chosen as an example design. This choice of design has two important benefits in terms of their (i) uniform size of sub-circuits that are in the output cone of each output, i.e., uniform size and connectivity density of networks that belong to longest paths and (ii) random internal connectivity. The 12-bit look-up table has been modeled in VHDL and synthesized using Synopsis and a subset of standard library only consisting of inverters and 2- and 3-input NAND and NOR gates. The reduction of the used cell library does not impact on the generality and the analysis could be easily conducted for circuits consisting of any type of gates. The full design consists of approximately 10^5 transistors.

After partitioning this large design using various partition sizes, sub-circuits that represent an output cone of each output in every partition are taken as sample circuits and sorted according to the logic depth of critical paths. Thus, each sample circuit has one output.

The probability of failure of sample circuits is directly acquired using the Monte Carlo (MC) tool described in Section 5.3. Since the tool provides values of the probability of failure for the worst-case logic-0 ($P^{\text{unit}}_{\text{fails},0}$) and logic-1 ($P^{\text{unit}}_{\text{fails},1}$), the probability of unit failure of a single output ($P^{\text{unit}}_{\text{fails}}$) is calculated using (7.5) from Section 7.1. The applied fault models assume permanent ("hard") faults that are constantly present in the system, and the probability of circuit failure is calculated as the worst case of all possible input vectors.

8.1.1 Dependency of Reliability on Logic Depth

Since the probability of failure of a circuit mainly depends on logic depth of its critical paths, the measure of this dependency using statistics of the probability of failure of large number of circuits for each logic depth need to be acquired. After acquiring the probability of failure of sample circuits using the MC tool, its statistics for each logic depth is evaluated and the mean value, the 95% confidence interval, and the upper bound are derived. More than 100 sample circuits have been evaluated for each logic depth.

The mean values as well as the bounds of the 95% confidence interval are given in Table 8.1 in units of the probability of failure of an equivalent gate ($P^{\text{gate}}_{\text{fails}}$). It is assumed that the equivalent gate consists of four transistors and that it fails for some fault types. On the other hand, some failures can be masked. Therefore, $P^{\text{gate}}_{\text{fails}}$ is proportional to the probability of individual device failure (p_{f}) and to the number of transistors, with an empirical coefficient k,

$$P^{\text{gate}}_{\text{fails}} = 4kp_{\text{f}}, \tag{8.1}$$

where $k = 0.2$ is a typical value for standard library gates (inverters and 2- and 3-input NAND and NOR gates) extracted using the MC tool (Section 5.3). This agrees well with the results presented in [31]. The results for logic depth values up to 15 are presented. For higher logic depths, the extrapolation based on the extracted

Table 8.1 The probability of circuit failure vs. logic depth (L)

Logic depth L	$P^{\text{unit}}_{\text{fails}}$	95% confidence interval for $P^{\text{unit}}_{\text{fails}}$
2	$2.25 \times P^{\text{gate}}_{\text{fails}}$	$[1.62,\ 2.88] \times P^{\text{gate}}_{\text{fails}}$
4	$5.67 \times P^{\text{gate}}_{\text{fails}}$	$[3.88,\ 7.46] \times P^{\text{gate}}_{\text{fails}}$
6	$11.63 \times P^{\text{gate}}_{\text{fails}}$	$[8.03,\ 15.23] \times P^{\text{gate}}_{\text{fails}}$
8	$20.86 \times P^{\text{gate}}_{\text{fails}}$	$[14.44,\ 27.28] \times P^{\text{gate}}_{\text{fails}}$
10	$38.56 \times P^{\text{gate}}_{\text{fails}}$	$[28.36,\ 48.76] \times P^{\text{gate}}_{\text{fails}}$
12	$59.32 \times P^{\text{gate}}_{\text{fails}}$	$[42.01,\ 76.63] \times P^{\text{gate}}_{\text{fails}}$
15	$117.89 \times P^{\text{gate}}_{\text{fails}}$	$[79.25,\ 156.53] \times P^{\text{gate}}_{\text{fails}}$

dependency is used, since circuits with logic depths higher than 15 are very large and impractical for statistical evaluation.

The dependence of the probability of failure on the logic depth is empirically demonstrated to be exponential and in the form given as

$$P_{\text{fails}}^{\text{unit}} = \sum_{i=1}^{L} F^{i-1} \cdot P_{\text{fails}}^{\text{gate}}, \tag{8.2}$$

where F is a parameter that is extracted through a fitting process and L is the logic depth of the circuit critical paths.

To understand the dependence expressed in (8.2), the tree model of the circuit with a single output is presented, which is illustrated in Fig. 8.4, where a tree structure of a circuit consisting of NAND gates is shown. Each NAND gate in the circuit has F inputs and the probability of failure of each gate is $P_{\text{fails}}^{\text{gate}}$. Therefore, F can be understood as the effective fanin of the gates. For example, if the effective fanin is 2, the total number of gates in the tree is $2^L - 1$, and the whole circuit is assumed to fail if any of the gates fail. Thus, $P_{\text{fails}}^{\text{unit}} = (2^L - 1) \cdot P_{\text{fails}}^{\text{gate}}$. However, in practice the tree structure of the circuit is not complete and has less than $(2^L - 1)$ gates. The upper bound of the probability of failure of the circuit is actually given by (8.2). Hence, it is assumed that every single output circuit can be represented in the format of this tree structure, and through the fitting process, the effective number of inputs that each gate in the equivalent tree circuit would have is extracted.

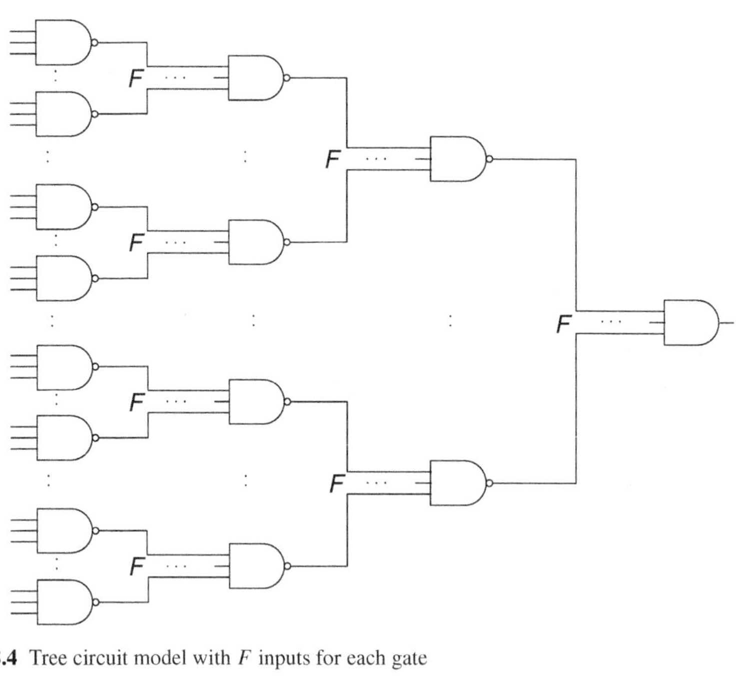

Fig. 8.4 Tree circuit model with F inputs for each gate

For each logic depth, the average value of the probability of circuit failure is extracted (depicted in Fig. 8.5), and the value of parameter F is numerically calculated. The following value is obtained (95% confidence parameter interval in brackets): $F = 1.33\,[1.24,\ 1.42]$. Since the worst case is examined, the upper bound value of 1.42 is taken as the value of the parameter. In Table 8.2, values of $P_{\text{fails}}^{\text{unit}}$ are calculated for higher logic depths, using (8.2) and the upper bound of the fitted parameter F.

Table 8.2 The probability of circuit failure vs. logic depth (L) for $L > 15$

Logic depth L	$P_{\text{fails}}^{\text{unit}}$
18	$388 \times P_{\text{fails}}^{\text{gate}}$
20	$1,170 \times P_{\text{fails}}^{\text{gate}}$
25	$5,688 \times P_{\text{fails}}^{\text{gate}}$
30	$31,770 \times P_{\text{fails}}^{\text{gate}}$
35	$182,360 \times P_{\text{fails}}^{\text{gate}}$
40	$1,051,800 \times P_{\text{fails}}^{\text{gate}}$

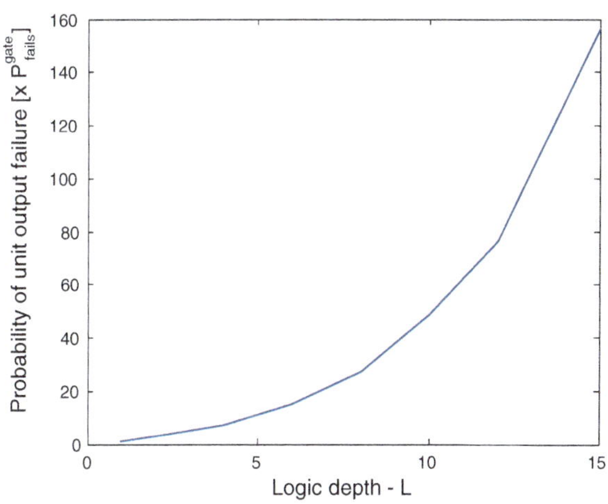

Fig. 8.5 Upper bound of probability of circuit failure vs. logic depth (L)

8.1.2 Reliability Improvement by Logic Depth Reduction

The fact that the probability of failure of a circuit depends on the logic depth of its critical paths can be exploited for redundancy-free local reliability optimization. The term redundancy free is used because no redundancy is applied in the circuit to achieve improvements in reliability.

In order to perform reliability improvement, a circuit can be synthesized in such a way that the logic depth of its critical paths is the minimal possible, and therefore its probability of failure is also reduced compared to non-optimal logic depth synthesis. To support these claims, an example using LGSynth'91 [251] benchmark circuit *b9* is evaluated with respect to its reliability, considering synthesized versions with different logic depths of critical paths, namely 7, 8, 9, and 10. *b9* is a mid-size benchmark circuit consisting of approximately 400 transistors, 41 inputs, and 21 outputs. The probability of failure is evaluated using the MC tool for all outputs and for p_f ranging from 0.001 to 0.01. Detailed MC simulations are used for better accuracy for $p_f = 0.005$. The values of $P_{\text{fails}}^{\text{unit}}$ for the most unreliable output and an average value over all outputs are reported in Table 8.3 for all four versions of *b9*.

Table 8.3 Probability of failure of the *b9* benchmark output vs. logic depth of the synthesized version for $p_f = 0.005$

Logic depth	Size	$P_{\text{fails}}^{\text{unit}}$ per output	
L	[num. eq. trans.]	Most unreliable	Average
7	424	0.134	0.065
8	384	0.135	0.069
9	354	0.115	0.072
10	388	0.121	0.079

The improvement in reliability between the versions with $L = 10$ and 7 equals 21.5% when the probability of failure is averaged over all outputs for $p_f = 0.005$. A constant improvement in reliability is noticeable with the reduction of the logic depth for all device probabilities of failure. For individual outputs, this is not necessarily the case, because the logic depth of the given output cone changes in different realizations. The average improvement in reliability for all device probabilities of failure averaged over all outputs is 18.8%. Realizations of the circuit with smaller logic depth have, in general, bigger sizes in terms of the number of equivalent transistors. The difference equals 16.5% between the smallest and the largest version.

The fact that reduction in logic depth improves reliability can be effectively used during the phase of reliability-optimal partitioning (Fig. 8.3), as will be demonstrated in Section 8.2.

8.1.3 Reliability Improvement of Different Fault-Tolerant Techniques

More significant reliability improvement can be achieved by using redundancy in different fault-tolerant techniques. The exact reliability improvement of different decision gates used in conjunction with R redundant units is explored in this section. The used decision gates are the following:

- Majority voter (MV) which allows correct operation even when $\frac{R-1}{2}$ out of R redundant units fail

- Averager with optimal fixed threshold (AVG), where a fixed threshold is set in order to minimize probability of failure for logic-0 and logic-1 simultaneously, as explained in Section 7.1
- Averager with adjustable threshold (4LRA), that together with redundant units, forms the full four-layer reliable architecture described in Section 6.1

In this section, a *fault-free* decision gate is assumed and the same samplecircuits as in the previous section are used as logic units in the analysis. Taking into consideration that redundant units and the decision gate are connected in series, and that their probabilities of failure can be assumed as independent, the evaluation of reliability of the full architecture can be performed by decomposing the reliable architecture into a series connection of the reliable architecture with a *fault-free* decision gate and a *faulty* decision gate. This is illustrated in Fig. 8.6 where the probability of failure of the reliable architecture with a *fault-free* decision gate is marked as $P_{\text{fails}}^{\text{red.unit}}$ and the probability of the decision gate failure is marked as $P_{\text{fails}}^{\text{dec.gate}}$.

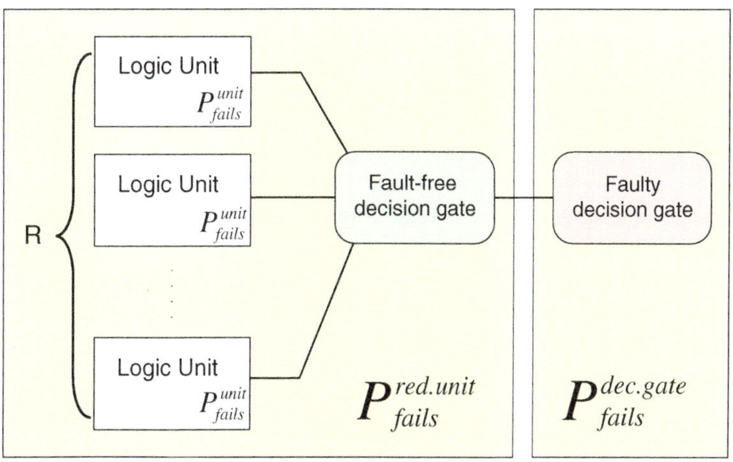

Fig. 8.6 Redundant units and *fault-free* decision gate in series connection with a *faulty* decision gate

Data for the following analysis are obtained by applying the statistical analysis procedure presented in Section 7.1 on PDFs extracted using the advanced single-pass reliability evaluation tool (presented in Section 7.2).

The comparison of reliable techniques is performed on the basis of the redundancy which is necessary to achieve the target $P_{\text{fails}}^{\text{red.unit}}$ considering different defect densities. An example depicted in Fig. 8.7 gives a minimum redundancy factor for each fault-tolerant technique necessary to achieve $P_{\text{fails}}^{\text{red.unit}} = 10^{-4}$ for redundant units of logic depth three and different $P_{\text{fails}}^{\text{gate}}$. Besides the clear advantage of 4LRA over AVG and especially MV, note that 4LRA and AVG can take all integer values as a redundancy factor, as opposed to MV that can only take odd values. The reduction

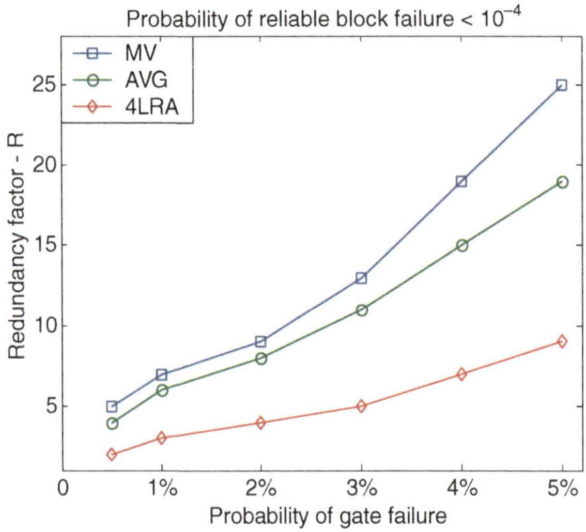

Fig. 8.7 Comparative analysis of necessary redundancy factor to keep the probability of reliable block failure smaller than 10^{-4} for 4LRA, AVG, and MV architectures plotted vs. the probability of gate failure

in the necessary redundancy factor reaches up to 25% for AVG and up to 65% for 4LRA.

In the following analysis, the reliability of MV, AVG, and 4LRA fault-tolerant techniques consisting of R units and a *fault-free* decision gate is evaluated with respect to the logic depth. In MV, a group of R units fails when at least $(R+1)/2$ units fail, where the probability of failure is following a binominal distribution. Therefore, following the discussion in Section 7.1 (Equations 7.9 and 7.10), $P_{\text{fails}}^{\text{red.unit}}$ is given as

$$\begin{aligned} P_{\text{fails}}^{\text{red.unit}} &= P_{\text{fails},0}^{\text{red.unit}} + P_{\text{fails},1}^{\text{red.unit}} - P_{\text{fails},01}^{\text{red.unit}} \\ &\approx 2 \cdot \sum_{i=\frac{R+1}{2}}^{R} \binom{R}{i} \left(P_{\text{fails}}^{\text{unit}}\right)^{i} \left(1 - P_{\text{fails}}^{\text{unit}}\right)^{R-i} \\ &\approx 2\binom{R}{\frac{R+1}{2}} \left(P_{\text{fails}}^{\text{unit}}\right)^{(R+1)/2} \left(1 - P_{\text{fails}}^{\text{unit}}\right)^{(R-3)/2} \left(1 - \frac{4P_{\text{fails}}^{\text{unit}}}{R+3}\right) \end{aligned} , \quad (8.3)$$

where $P_{\text{fails},0}^{\text{red.unit}}$ and $P_{\text{fails},1}^{\text{red.unit}}$ are probabilities of failure of the reliable architecture with a *fault-free* decision gate for the worst-case logic-0 and logic-1, respectively, $P_{\text{fails},01}^{\text{red.unit}}$ is the probability of simultaneous failure for logic-0 and logic-1 values (see Section 7.1), and $P_{\text{fails}}^{\text{unit}}$ represents the highest value of the probability of failure of a unit output between logic-0 and logic-1. In (8.3), following assumptions have been made:

- $P^{\text{red.unit}}_{\text{fails},0/1} < 1\%$, $P^{\text{red.unit}}_{\text{fails},01} / P^{\text{red.unit}}_{\text{fails},0/1} < 1\%$ and
- $P^{\text{unit}}_{\text{fails}} = \max(P^{\text{unit}}_{\text{fails},0}, P^{\text{unit}}_{\text{fails},1}) < 10\%$.

The second assumption is justified since $P^{\text{red.unit}}_{\text{fails},01} \sim \left(P^{\text{red.unit}}_{\text{fails},0/1}\right)^2$. The upper bound for the relative error in (8.3) is 1% since both approximations have the opposite impact on the absolute value.

If i units have failed in advance from a different cause (e.g., input signals failure), the probability of failure of the reliable architecture with a *fault-free* decision gate becomes the probability that at least $\frac{R+1}{2} - i$ units will fail out of $R - i$ units

$$P^{\text{red.unit}}_{\text{fails},(i)} \approx 2\binom{R-i}{\frac{R+1}{2}-i}(P^{\text{unit}}_{\text{fails}})^{\frac{R+1}{2}-i} \cdot (1 - P^{\text{unit}}_{\text{fails}})^{(R-3)/2}\left(1 - \frac{4-2i}{R+3-2i}P^{\text{unit}}_{\text{fails}}\right). \tag{8.4}$$

A straightforward formula given in (8.5) and derived using Stirling's approximation $n! \approx \sqrt{2\pi n}\,(n/e)^n$ provides an accurate estimation of binominal coefficient $\binom{R}{\frac{R+1}{2}}$ even for low values of R ($R < 10$). In Table 8.4, exact values obtained by direct calculation of the binominal coefficient values obtained using our approximation formula (8.5) and relative errors between these two are given:

$$\binom{R}{\frac{R+1}{2}} = \frac{4^{(R+1)/2}}{\sqrt{2\pi(R+1.5)}}. \tag{8.5}$$

The relative error is only 0.3%, in the worst case for $R = 3$. When increasing R, the relative error reduces.

Table 8.4 Binominal coefficient estimation for various redundancy factors (R)

$\binom{R}{\frac{R+1}{2}}$	$R = 3$	$R = 5$	$R = 7$	$R = 9$	$R = 99$
Exact value	3	10	35	126	5.04460×10^{28}
Our approx.	3.009	10.015	35.03	126.07	5.04457×10^{28}
Rel. err. (%)	0.3	0.146	0.086	0.056	0.0006

Finally, the probability of block failure with a *fault-free* decision gate is given as

$$P^{\text{red.unit}}_{\text{fails}} \approx 2\frac{(1 - P^{\text{unit}}_{\text{fails}})^{(R-3)/2}}{\sqrt{2\pi(R+1.5)}}\left(1 - \frac{4P^{\text{unit}}_{\text{fails}}}{R+3}\right)(4P^{\text{unit}}_{\text{fails}})^{(R+1)/2} \approx A(4P^{\text{unit}}_{\text{fails}})^{E_1(R+1)/2}, \tag{8.6}$$

where

$$A = \frac{(1 - P_{\text{fails}}^{\text{unit}})^{(R-3)/2}}{\sqrt{2\pi(R+1.5)}} \left(1 - \frac{4P_{\text{fails}}^{\text{unit}}}{R+3}\right) \quad \text{and} \quad E_1 = 1.$$

The exponential factor E_1 is taken as the reliability improvement factor.

For AVG and 4LRA, the probability of failure of the reliable block with a *fault-free* decision gate cannot be expressed in an analytical form. The dependence between $P_{\text{fails}}^{\text{red.unit}}$ and the redundancy factor for AVG and 4LRA (assuming that the same expression is valid) is similar, but with the difference in the exponential factor ($E_{2/3}$) for AVG and 4LRA, respectively. For MV, the reliability improvement factor (E_1) does not depend on the size of the unit (logic depth) or $P_{\text{fails}}^{\text{unit}}$. For AVG and 4LRA on the other hand, $E_{2/3}$ depends on $P_{\text{fails}}^{\text{unit}}$ and subsequently on the logic depth because the unit PDF depends on $P_{\text{fails}}^{\text{unit}}$, as established in Section 7.2. Here, the dependence of $E_{2/3}$ on logic depth (L) is determined. Following the values of $P_{\text{fails}}^{\text{unit}}$ for different logic depths from Table 8.1, PDFs of the corresponding probabilities of unit failure have been generated according to the method presented in Section 7.2. In a subsequent step, these PDFs are evaluated as explained in Section 7.1. Finally, the acquired values of $P_{\text{fails}}^{\text{red.unit}}$ are fitted according to (8.6) to acquire the $E_{2/3}$ parameters. Linear fitting is applied to the logarithm of curves expressed (8.6). The results are presented in Table 8.5 with the mean values, and the bounds, for the 95% confidence interval. In Fig. 8.8, the dependence between $P_{\text{fails}}^{\text{red.unit}}$ and R is depicted for $L = 3$. Even though only odd values of R are depicted, the exponential factor has been fitted using odd and even values of R, $R \in [2, 25]$. A clear exponential dependence between $P_{\text{fails}}^{\text{red.unit}}$ and the redundancy factor R for AVG/4LRA as well as for MV is observed. This justifies the assumption that the dependence is the same as the one given in (8.6) for MV.

Table 8.5 Dependence of the exponential factor on logic depth for AVG and 4LRA

Logic depth L	E_2	95% confidence interval for E_2	E_3	95% confidence interval for E_3
2	1.047	[1.041, 1.053]	2.016	[1.997, 2.036]
3	1.018	[1.016, 1.021]	1.942	[1.922, 1.962]
4	1.009	[1.008, 1.011]	1.921	[1.905, 1.937]
5	1.005	[1.004, 1.006]	1.911	[1.901, 1.921]
6	1.003	[1.002, 1.004]	1.905	[1.898, 1.912]
10	1.001	[1.000, 1.002]	1.901	[1.897, 1.905]

Practically, for $L > 5$, E_2 becomes constant with respect to logic depth (and subsequently to $P_{\text{fails}}^{\text{unit}}$) and equal to MV. This is at the same time, the point up to which AVG shows improvement compared to MV. Regarding AVG, E_3 becomes practically constant for $L > 6$. However, the exponential factor $E_3 = 1.9$ suggests that the improvement in reliability of 4LRA compared to MV and AVG is almost quadratic, which provides potential to achieve the same reliability with 45% reduced redundancy factor.

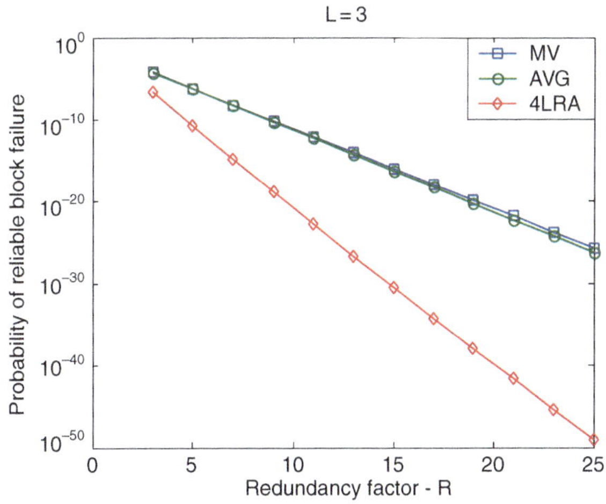

Fig. 8.8 Comparative analysis of 4LRA, AVG, and MV in terms of probability of failure of the reliable block with a *fault-free* decision gate for different redundancy factors

Similarly as in Section 8.1.1, the probability of failure is analyzed for four different versions of the *b9* circuit used as a redundant unit comprising fault-tolerant architectures. The different versions have been synthesized targeting four logic depths of their critical paths, namely 7, 8, 9, and 10. The possible improvement in reliability is obtained as a benefit of two factors:

- reliability improvement factor (E) and
- reduced probability of redundant unit output failure for smaller logic depths.

The probability of failure is evaluated using the MC tool (Section 5.3) for all outputs and for a device probability of failure p_f ranging from 0.001 to 0.01. Detailed MC simulations are used for better accuracy for $p_f = 0.005$. The values of probability of failure for the most unreliable output and an average value over all outputs are reported in Table 8.6 for all four logic depth versions of *b9*, two redundancy factors ($R = 3$ and $R = 5$), and three fault-tolerant architectures (MV, AVG, and 4LRA). The values are evaluated following (7.10), (7.15), and (7.18) for MV, AVG, and 4LRA respectively.

The improvement in reliability between the versions comprising $L = 10$ and 7 equals 47.8, 48.2, and 84.6% in the average case over all outputs for $R = 3$ and MV, AVG, and 4LRA fault-tolerant configurations, respectively. For $R = 5$ the improvement is even larger, i.e., 51, 55, and 92% in the average case over all outputs for MV, AVG, and 4LRA fault-tolerant configurations, respectively. A constant improvement in reliability is noticeable with the reduction of the logic depth for all device probabilities of failure, when the average probability of failure over all outputs is taken into consideration. For individual outputs, this is not the case, because the logic depth of the output cone changes in different realizations. The

Table 8.6 Probability of failure of the *b9* benchmark output vs. logic depth of the synthesized version for $p_f = 0.005$, for redundancy factors equal to 3 and 5, and MV, AVG, and 4LRA fault-tolerant techniques

Logic depth L	Fault-tol. technique	$R = 3: P_{\text{fails}}^{\text{unit}}$ per output Most unreliable	$R = 3$: Average	$R = 5: P_{\text{fails}}^{\text{unit}}$ per output Most unreliable	$R = 5$: Average
	MV	0.00862	0.00276	1.18×10^{-3}	3.33×10^{-4}
7	AVG	0.00821	0.00263	1.08×10^{-3}	2.96×10^{-4}
	4LRA	0.000684	0.000195	5.99×10^{-5}	1.63×10^{-5}
	MV	0.0102	0.00334	1.358×10^{-3}	3.92×10^{-4}
8	AVG	0.00972	0.00321	1.246×10^{-3}	3.51×10^{-4}
	4LRA	0.000896	0.00028	6.39×10^{-5}	1.91×10^{-5}
	MV	0.00883	0.00377	1.21×10^{-3}	4.38×10^{-4}
9	AVG	0.00841	0.00349	1.11×10^{-3}	3.96×10^{-4}
	4LRA	0.000746	0.00031	6.08×10^{-5}	2.46×10^{-5}
	MV	0.00961	0.00408	1.31×10^{-3}	5.03×10^{-4}
10	AVG	0.00915	0.00390	1.22×10^{-3}	4.59×10^{-4}
	4LRA	0.000825	0.00036	6.77×10^{-5}	3.13×10^{-5}

improvement is significant for all three fault-tolerant techniques compared to the case where no redundancy is applied (Section 8.1.1) and remains constant in the explored defect density range $p_f \in [0.001,\ 0.01]$.

The presented results demonstrate that local optimization can be efficiently performed by only resynthesizing circuits, having the logic depth as the minimization goal. This will be further exploited in the following section.

In this section an estimation of the probability of failure of small to mid-sized circuits has been given with respect to the logic depth of these circuits. The evaluation of the effectiveness of analog averaging techniques in terms of reliability improvement compared to majority voter-based techniques has been performed and it has been proven that the effectiveness significantly reduces with increased logic depth. Moreover, it has been shown that the local optimization can be efficiently performed by resynthesizing circuits, having the logic depth as the minimization goal.

Considering the importance of the decision gate reliability in the overall system reliability, this aspect will be explored in detail in Section 8.3, having in mind that the realization of the adaptive threshold in 4LRA is costly in terms of size compared to AVG or MV. Since the probability of failure of the decision gate is proportional to its size, the advantage of 4LRA compared to AVG and MV that is demonstrated in this section is significantly reduced when relatively small redundant units and fault-prone decision gates are included.

8.2 Optimal Reliability Partitioning

Circuit partitioning consists of dividing the circuit into parts, each of which can be implemented as a separate component (e.g., a chip), that satisfies demanded design constraints. The partitioning of the system is an important step in the system-level optimization that hierarchically divides a system into local and system levels. The emphasis is on the problem of combinational circuit partitioning with the goal of minimizing probability of failure minimization subject to area constraints.

The total chip probability of failure depends on various factors, as discussed in detail in Section 8.3:

- number of partitions,
- probability of failure of each output of a partition,
- average number of outputs of partitions, and
- average number of inputs of the partitions (for some fault-tolerant techniques).

As it is shown in the previous section, the probability of failure of each output of the partition largely depends on the circuit logic depth. Therefore, a minimizing probability of failure constraint can be substituted to a minimizing logic depth constraint, which is equivalent to a minimizing circuit delay constraint, where the reliability weighting factor is attached to each net in replacement of a delay parameter. Minimizing the average number of outputs and inputs of partitions is equivalent to minimizing cut sizes [252].

Partitioning approaches that have the minimization of the circuit delay as an important constraint belong to a class of so-called time-driven partitioning approaches [253–257] and can be classified into two categories: (i) *top-down* partitioning approaches and (ii) *bottom-up* clustering-based approaches. Approaches in the first category are usually based on the Fiduccia–Mattheyses (FM) [258] recursive min-cut partitioning method or on quadratic programming formulations [259, 260]. Timing optimization is obtained by minimizing the delay of the most critical path. The approaches from the second category are mostly used as a preprocessing step for min-cut algorithms [261–263]. All previous approaches achieve delay minimization by netlist alteration such as logic replication, retiming, and buffer insertion in order to meet delay constraints while the cut size is minimized. The focus is on delay improvement, and the cut size is ignored. Gate replication can be massive in these methods.

Most of these approaches are not suitable to be applied for reliability minimization partitioning, and reliability cannot be improved by using techniques such as logic replication, retiming, and buffer insertion. Moreover, the runtime for moderate-sized circuits is excessive and makes these approaches impracticable for large-sized circuits. One reason for that may be that previous approaches usually separate the timing-driven partitioning into two steps: (i) clustering or partitioning and (ii) timing refinement based on netlist alteration [259, 261].

The approach adopted in this book attempts to eliminate the above deficiencies by assessing timing (reliability)-driven partitioning from a different perspective: the probability of failure of each net of the circuit which is acquired by the modified single-pass reliability evaluation tool (Section 7.2) is used to change the partitioning process itself to perform minimization of probability of failure. The very fast hMetis partitioning algorithm [250, 252] is used.

The novel adopted partitioning approach as well as existing approaches has common practical limitations related to the total size of the circuit to be partitioned and the size of partitions. All existing approaches become inefficient in terms of cut size, and runtime becomes impractically large from a certain break-even circuit/netlist size. Therefore, a different approach to partitioning is necessary for partitioning of very large design, e.g., functional partitioning. Moreover, regular design usage should be encouraged to help partitioning and improve reliability.

Two distinctive cases are assessed based on the target partition size:

- small to mid-sized partitions (less than 10^5 devices) and
- large-sized partitions (over 10^5 devices). In the first case, an efficient reliability-driven procedure described above is provided.

For the first case, the full solution is proposed.

8.2.1 Partitioning to Small and Mid-Sized Partitions

A multi-objective partitioning scheme that is performing a simultaneous cut size and probability of failure minimization is presented. The partitioning is done by recursive bipartitioning [250, 252]. At each level, a reliability factor of the net is associated as a weight with all corresponding hyperedges in the hypergraph. Then, the hMetis partitioning algorithm is run using the hyperedge coarsening scheme. In this scheme, during hypergraph coarsening, the hyperdges that have large weights are less prone to be cut by the partitioner. By using the reliability factor as a hyperedge weight, the edges that will be cut can be efficiently controlled. Reliability factors (i.e., hyperedge weights) are updated at each partitioning level. Initially, all reliability factors in the circuit are computed using the modified single-pass reliability analysis tool. These reliability factors are then used as weights associated with hyperedges. After the first bipartitioning, the reliability of each partition is reevaluated; new reliability factors are generated and attached to hyperedges as weights. During the recursive bipartitioning new reliability factors are generated in each step. The recursive bipartitioning process stops when each block contains a number of vertices which is smaller than a specified threshold. The pseudo-code of the proposed reliability-driven hMetis-based partitioning algorithm is given in Algorithm 1.

Important functions of the algorithm `compute_reliability()` and `assign__weights()` are explained using the example shown in Fig. 8.9. The example circuit is ISCAS-C17 [249]. In `compute_reliability()`, the modified single-pass reliability analysis is performed and the probability of failure of each net related to the output of each gate is calculated (depicted as a red value, at the left of the brackets, attached to each hyperedge in Fig. 8.9b) where the hypergraph is shown as a directed acyclic graph (DAG)). The used gate probability of failure is 10%. The `assign_weights()` function takes the probability of failure value of each hyperedge ($P_{f,i}$) and assigns the weight to the hyperedge (W_i, depicted as

Algorithm 1 Reliability-driven partitioning.

Goal: Partition a circuit into reliability optimal partitions

```
Queue = G(V,E); // queue initialized with initial graph
compute_reliability(G);
assign_weights(G);
while (Queue not empty)
  g=pop(Queue);
  (gA, gB) = partition_with_hMetis(g);
  if (size_of(gA)>T) // T = max # of gates per partition
    push(Queue, gA);
  endif
  if (size_of(gB)>T)
    push(Queue, gB);
  endif
  compute_reliablity(gA,gB);
  assign_weights(gA,gB);
end
```

blue value, inside brackets attached to each hyperedge in Fig. 8.9b) according to the function given as follows:

$$W_i = \begin{cases} \left(\frac{1}{P_{\mathrm{f,max}}}\right)^{\alpha} & \text{for } P_{\mathrm{f},i} = 0 \\ \left(\frac{(P_{\mathrm{f,max}})^2}{P_{\mathrm{f},i}}\right)^{\alpha} & \text{for } 0 < P_{\mathrm{f},i} \leq (P_{\mathrm{f,max}})^2 \, , \\ \left(\frac{P_{\mathrm{f},i}}{(P_{\mathrm{f,max}})^2}\right)^{\alpha} & \text{for } P_{\mathrm{f},i} > (P_{\mathrm{f,max}})^2 \end{cases} \tag{8.7}$$

where α is the optimization parameter and $P_{\mathrm{f,max}}$ is the probability of failure of the output of the cone to which the observed gate belongs to. For example, gate G_1 in Fig. 8.9a belongs to the cone whose output is also the output of the gate G_5 (marked as 11 in Fig. 8.9b). If a gate belongs to multiple cones (like G_3 in Fig. 8.9a) then $P_{\mathrm{f,max}}$ is assigned with the largest value of W_i for all the cones that the observed gate belongs to (output marked as 12 in Fig. 8.9b is taken for $P_{\mathrm{f,max}}$ for gate G_3). The function given in (8.7) tends to assign smaller weights to the nets that are positioned far from inputs and outputs. This way, a higher priority is given to those nets to be cut by the partitioner and at the same time a higher priority is given to the nets which are located closer to the beginning and the end of critical paths to be clustered together in the hyperedge coarsening scheme of the hMetis partitioner [252]. The optimization parameter α is used to provide an optimal balance between minimizing edge-cuts and minimizing the probability of failure. For $\alpha = 0$ all weights of hyperedges are assigned to one and the partitioning is identical to the pure hMetis partitioning. With the increase of α, the importance of minimization of the probability of failure is increased.

The partitioning is applied to the 12-bit look-up table design (described in Section 8.1) partitioning the design into partitions of a given size. The statistics on partition's number of inputs (fanin F_{in}), number of outputs (fanout F_{out}), and logic

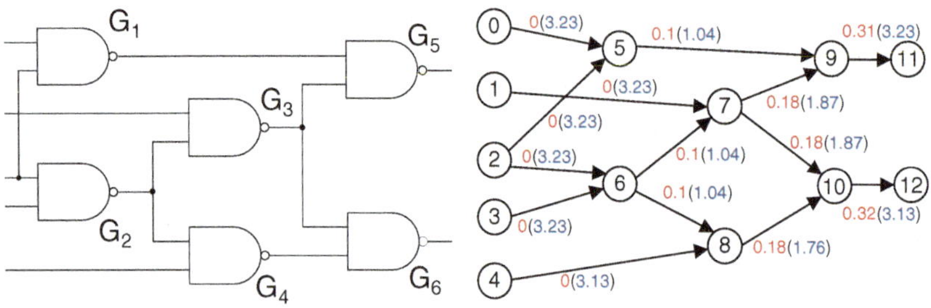

(a) (b)

Fig. 8.9 (**a**) Example circuit for partitioning and (**b**) hypergraph of the example circuit for partitioning with weights

depth (L) have been made. The average values of F_{in}, F_{out}, and upper bound (95% confidence interval) for L are given in Table 8.7.

Table 8.7 Partitioning statistics of F_{in}, F_{out}, and L for different partition sizes

Partition size N_c	Fanin F_{in}	Fanout F_{out}	Logic depth L
10	3	1	≤ 2
32	8	3	≤ 3
100	24	8	≤ 5
320	40	15	≤ 7
1,000	96	36	≤ 9
3,200	241	90	≤ 11
10,000	543	203	≤ 13
32,000	1,185	456	≤ 15

8.2.2 Partitioning to Large-Sized Partitions

There are no demonstrated examples in the literature of designs that have a target partition size (N_c) larger than 10^5 devices, and thus it can be assumed that cut size minimization approaches are inefficient for large-sized partitions, and a functional partitioning is necessary. Conceptual values of hypothetical functional partitioning of large designs are given here, since they are used in further analyses in Section 8.3.

The partitioning is assumed to group the system-level partitions' input/output signals into a common bus of width B, for the whole chip (a hypothetical partitioned design is depicted in Fig. 8.10). The selected values of B correspond to values that are common in today's processing cores. Two different values of B are assumed:

- $B = 128$ for $N_c \in [10^5, 10^7]$ (typical value for today's chips of a similar size) and
- $B = 256$ for $N_c \in [10^7, 10^9]$ (near future projected value).

It is also assumed that values of logic depth with respect to the size of partitions correspond to values which are common in nowadays processing cores [264]. These values are given in Table 8.8. The actual optimal reliability functional partitioning is out of the scope of this book.

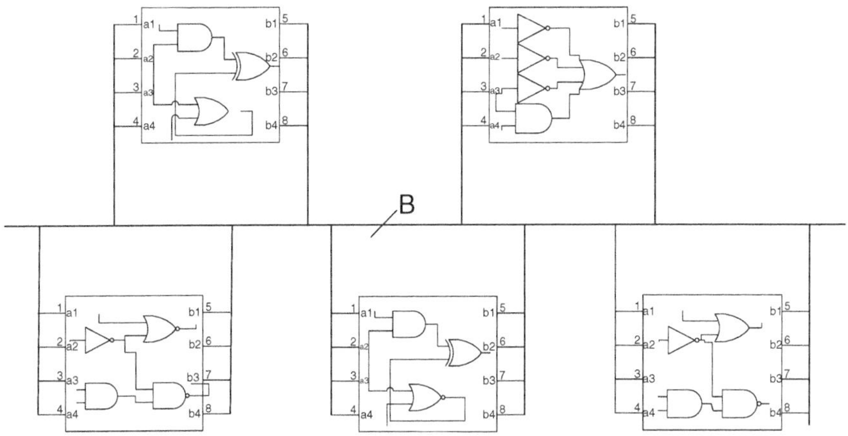

Fig. 8.10 Example of functional partitioning of a large design into partitions where all partition inputs and outputs are part of the same bus

Table 8.8 Logic depth for different partition size for $N_c \geq 10^5$

Partition size N_c	Logic depth L
1.0×10^5	≤ 18
3.2×10^5	≤ 21
1.0×10^6	≤ 23
3.2×10^6	≤ 25
1.0×10^7	≤ 28
3.2×10^7	≤ 31
1.0×10^8	≤ 35
3.2×10^8	≤ 38
1.0×10^9	≤ 41

8.3 System-Level Evaluation and Optimization

In order to design a system consisting of a large number of unreliable devices, a strategy would first consist of partitioning the entire system into reliability optimally sized partitions, taking each partition as a unit, applying one reliability technique on that unit to build a reliable block, and then optimally combining these reliable blocks according to various fault-tolerant techniques. System optimization can be performed according to various design parameters. However, two parameters, namely the partition size (N_c) and the reliable block redundancy factor (R), have the highest importance in the optimization process. The partitioning has to be as uniform as possible in terms of size and logic depth. Uniform partitions can be assumed to have a reliability within acceptable boundaries which are defined in Section 8.1. With uniform partitioning, the optimal partition sizes and redundancy factors are assumed to be the same for all the partitions.

Four highly generic fault-tolerant techniques using redundancy have been analyzed in terms of reliability evaluation and optimization:

- R-fold modular redundancy with a decision gate (RMR) (Fig. 8.11a) is a generalization of the TMR configuration where R units operate in parallel, and $R = 3, 5, 7, 9, \ldots$, and a decision is needed to build this configuration. There are three possible configurations depending on the decision gate implementation:

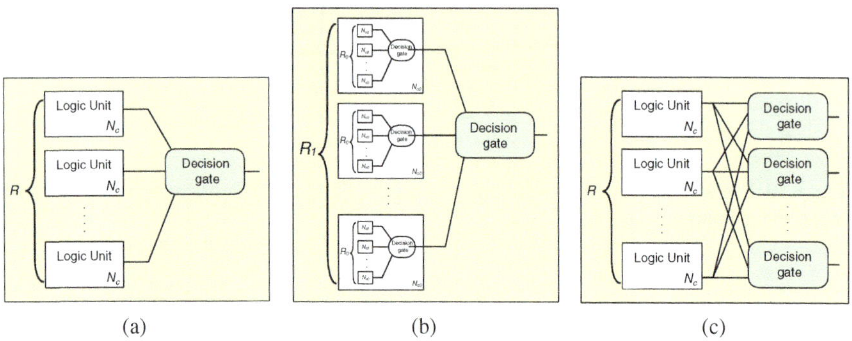

Fig. 8.11 (**a**) RMR; (**b**) CRMR; and (**c**) DRMR

  - R-fold modular redundancy with a majority voter (RMR-MV), where the decision gate is a majority voter implementing $(R + 1)/2$ out of R majority function,
  - R-fold modular redundancy with an averager (RMR-AVG), where the decision gate is an averager with the fixed optimal threshold, and
  - R-fold modular redundancy in a 4LRA configuration (RMR-4LRA), where the decision gate is an averager with an adaptive threshold forming together with R units a 4LRA.

- Cascaded R-fold modular redundancy (CRMR) (Fig. 8.11b) is a concept similar to RMR, in which the units working in parallel are RMR units combined with a decision gate. This configuration forms a "first-order" CRMR. RMR can be considered to be "zeroth-order" CRMR. Any order of cascading can be considered; however, the reliability of the final system does not necessarily increase with the cascading order number. Similarly as for RMR, three configurations are possible depending on the decision gate implementation:

  - cascaded R-fold modular redundancy with a majority voter (CRMR-MV),
  - cascaded R-fold modular redundancy with an averager (CRMR-AVG), and

  - cascaded R-fold modular redundancy in a 4LRA configuration (CRMR-4LRA).

- Distributed R-fold modular redundancy (DRMR) (Fig. 8.11c) is a concept similar to RMR, where each output is connected to several decision gates. The idea behind the concept is to increase the reliability of decision gates that are perceived as critical for fault tolerance in RMR, which will be further demonstrated. N decision gates can operate in parallel, and $N = 1, 2, \ldots, R$. Theoretically the number of decision gates can be higher than the number of redundant units ($N > R$). However, this realization has no practical advantage since decision gates are feeding their outputs to R units, and when $N > R$ further collapsing of surplus signals is necessary through the second layer of decision gates that are completely negating the benefit of additional ($N - R$) decision gates in the first layer. For $N < R$ the efficiency of DRMR in terms of fault tolerance is reduced (RMR is a special case of DRMR for $N = 1$) since there is an exponential decrease in reliability improvement and only linear reduction of the overhead. Since the goal is the maximization of the reliability, $N = R$ is taken as an optimal case.

  - distributed R-fold modular redundancy with a majority voter (DRMR-MV),
  - distributed R-fold modular redundancy with an averager (DRMR-AVG), and
  - distributed R-fold modular redundancy in a 4LRA configuration (DRMR-4LRA).

- NAND multiplexing; von Neumann multiplexing realized as parallel restitution [123].

The common evaluation and optimization procedure proposed in the introduction of this chapter can be independently applied to RMR, CRMR, and DRMR fault-tolerant techniques.

All decision gate types have been used in the RMR/CRMR/DRMR analysis even though, as presented in Section 8.1, the reliability improvement of AVG and 4LRA techniques depends on the logic depth, i.e., the size of a unit. When the logic depth of the optimal size of partitions is larger than the values suggested in Section 8.1, the exponential factor for AVG and 4LRA techniques ($E_{2/3}$) is assumed to be constant.

Each of these techniques is analyzed separately, in detail, and for each of them, the optimization procedure for acquiring global parameters such as the optimal partition size (granularity level) (illustrated for two different sizes in Fig. 8.12) and the optimal reliable block redundancy factor is given in the following sections. For each of the techniques, the analysis and the optimization procedure are demonstrated in the design example where a large hypothetical design (referred in further discussions as chip) is used. Finally, the comparison of all the techniques as well as the optimal range of application is subsequently derived and discussed in Section 8.3.5.

The following estimations and assumptions in the analysis of the RMR, CRMR, and DRMR have been adopted to reflect consistent working hypothesis, and the analysis is extended, also considering the AVG and 4LRA:

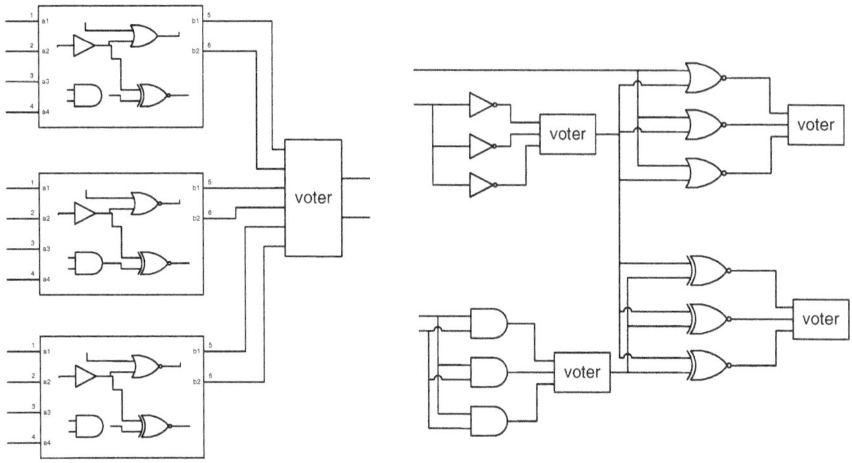

Bigger size of fault-tolerant clusters – finer granularity

Smaller size of fault-tolerant clusters – coarser granularity

Fig. 8.12 Different size of fault-tolerant partitions, with identical functionality

1. In [6], the total number of devices on a chip (N_{tot}) is kept constant and the redundancy factor is optimized to obtain the best fault tolerance. The drawback of the approach is related to the increased redundancy which reduces the number of functions which the chip can realize. Instead of keeping the total number of devices on a chip constant, the functionality of a chip is guaranteed, regardless of the applied redundancy factor or fault-tolerant cluster size. The number of devices that guarantee the functionality is in the further text referred to as the effective number of devices (N).
2. Only moderate redundancy factors ($R < 1,000$) are regarded as feasible. Increased redundancy increases the total overhead. Considering the ultimate device density of 10^{12} devices per cm^2, having a chip with 10^{12} devices in total limits the effective number of devices to 10^9, where a maximal overhead is considered to be equal to 1,000. Thus, the optimization goal is to develop fault-tolerant techniques which enable correct functioning of a chip consisting of 10^9 effective devices with the probability of 90%.
3. The number of necessary devices to realize decision gates (majority/averaging/thresholding) for each output depends on the number of inputs of the voter gate (e.g., the number of redundant units), and a linear dependency on the redundancy factor R with the number of inputs is assumed. This assumption is more realistic than the assumption made in [6], where a constant voter gate size is used, for various redundancy factors.
4. The number of decision gates is equal to the number of outputs of each unit that a decision gate is processing (fanout F_{out}). Moreover, the probability of failure of the reliable block is assumed to be the sum of probabilities of failure for each

output of the block. The probability of failure of the reliable block also depends on the number of inputs of each unit (fanin F_{in}) for the DRMR fault-tolerant technique as shown in Section 8.3.3. The fanin and fanout are nonlinear functions of a partition size (N_c) as could be seen from Table 8.7 and two different estimations based on the optimal size of a partition and target design size are provided:

a. For optimal partition sizes smaller than 10^5 devices, the values of fanin and fanout are given in Table 8.7.
b. For optimal partition sizes larger than 10^5 devices, functional partitioning is assumed (Section 8.2.2), which imposes that all signals between partitions are grouped into buses of a width equal to B. Therefore, $B = F_{\text{in}} = F_{\text{out}}$, and the selected value corresponds to processing units of similar size in today's processing cores. Two different values of B are assumed:

   - $B = 128$ for $N_c \in [10^5,\ 10^7]$ and
   - $B = 256$ for $N_c \in [10^7,\ 10^9]$.

5. Decision gates are assumed to consist of $(m_1 + 2R)$, $(m_2 + 2R)$, and $(m_3 + 2R)$ devices for the MV, AVG, and 4LRA, respectively, and to have an input stage formed with two transistors per input, and an output stage which is performing the majority function, averaging and fixed/adaptive thresholding. $m_{1/2/3}$ is the number of transistors used to realize the output stage performing the majority function, averaging with fixed and averaging with adaptive thresholding, respectively. For the majority function a static "mirrored adder" configuration [229] and $m_1 = 10$ are assumed. A possible smaller configuration is presented in [265]. For averaging with fixed thresholding, a CMOS floating gate realization [228] with the output buffer and $m_2 = 4$ are considered. For averaging with adaptive thresholding, a configuration that exploits the transistor as a four-terminal device [141, 229] is considered. An additional circuitry that is realizing adaptive thresholding as explained in Section 6.1 and $m_3 = 100$ are also assumed. The probability of a decision gate failure ($P_{\text{fails}}^{\text{dec.gate}}$) is assumed to be proportional to gate's number of devices, i.e.

$$P_{\text{fails}}^{\text{dec.gate}} = k(m_{1/2/3} + 2R)p_{\text{f}}, \tag{8.8}$$

where k is the factor of proportionality and p_{f} is the probability of individual device failure. As a critical component for the reliability, a decision gate can be realized with more reliable devices than the rest of the circuit yielding $k \ll 1$. In analyses presented in this section, it is assumed that $k = 0.2$, which is a typical value for standard library gates extracted using the MC tool (Section 5.3).
6. The probability of unit failure $P_{\text{fails}}^{\text{unit}}$ depends on the logic depth and is given in Table 8.1. The relation between the logic depth (L) and the partition size (N_c) ranges is given in Tables 8.7 and 8.8. Combining these two tables and replacing $P_{\text{fails}}^{\text{gate}}$ with $0.8p_{\text{f}}$ (see (8.1)) yield the relation between $P_{\text{fails}}^{\text{unit}}$ and N_c that is

Table 8.9 Probability of unit output failure for different partition sizes

Partition size N_c	P_{fails}^{unit}	Partition size N_c	P_{fails}^{unit}
10	$1.13p_f$	1.0×10^5	$525p_f$
32	$2.75p_f$	3.2×10^5	$1{,}410p_f$
100	$6p_f$	1.0×10^6	$3{,}200p_f$
320	$12.5p_f$	3.2×10^6	$6{,}820p_f$
1,000	$25.8p_f$	1.0×10^7	$1.71 \times 10^4 p_f$
3,200	$52.5p_f$	3.2×10^7	$4.68 \times 10^4 p_f$
10,000	$106p_f$	1.0×10^8	$1.67 \times 10^5 p_f$
32,000	$215p_f$	3.2×10^8	$5.12 \times 10^5 p_f$
		1.0×10^9	$1.5 \times 10^6 p_f$

provided in Table 8.9. When combining the tables, upper bound values for P_{fails}^{unit} are taken.

7. The probability of failure of redundant units (reliable block failure with *fault-free* decision gates), according to the analysis presented in Section 8.1 is expressed as

$$\begin{aligned} P_{fails}^{red.unit} &\approx 2\binom{R}{\frac{R+1}{2}}(1 - P_{fails}^{unit})^{(R-3)/2} \\ &\quad \cdot \left(1 - \frac{4P_{fails}^{unit}}{R+3}\right)(P_{fails}^{unit})^{E_{1/2/3}(R+1)/2} \\ &\approx 2\frac{(1 - P_{fails}^{unit})^{(R-3)/2}}{\sqrt{2\pi(R+1.5)}}\left(1 - \frac{4P_{fails}^{unit}}{R+3}\right)(4P_{fails}^{unit})^{E_{1/2/3}(R+1)/2} \\ &\approx A(4P_{fails}^{unit})^{E_{1/2/3}(R+1)/2} \end{aligned}, \quad (8.9)$$

where

$$A = 2\frac{(1 - P_{fails}^{unit})^{(R-3)/2}}{\sqrt{2\pi(R+1.5)}}\left(1 - \frac{4P_{fails}^{unit}}{R+3}\right)$$

and $E_{1/2/3}$ is the exponential factor that depends on the logic depth and is given in Table 8.5. Combining this table with Tables 8.7 and 8.8 yields the relation between $E_{1/2/3}$ and N_c that is given in Table 8.10. Indices 1, 2, and 3 correspond to MV, AVG, and 4LRA, respectively.

Table 8.10 Exponential factor for AVG and 4LRA decision gates for different partition sizes

Partition size N_c	E_2	E_3
10	1.047	2.016
32	1.018	1.942
100	1.005	1.911
> 320	1	1.9

8.3.1 R*-Fold Modular Redundancy (RMR)*

Depending on the partition size (N_c), $N_p = N/N_c$ partitions need to be replicated. Therefore, there will be N_p reliable blocks in the entire chip.

A reliable block with an *imperfect* decision gate can be evaluated in terms of reliability as a series connection of a reliable block with a *fault-free* decision gate (8.9) and an imperfect decision gate (8.8) and is expected to fail if any element in the series connection fails. This yields the probability of failure of one output of the reliable block with *imperfect* decision gate ($P_{\text{fails}}^{\text{block}}$) as the sum of probabilities of failure given in (8.9) and (8.8),

$$\begin{aligned} P_{\text{fails}}^{\text{block}} &\approx P_{\text{fails}}^{\text{red.unit}} + P_{\text{fails}}^{\text{dec.gate}} \\ &\approx A(4P_{\text{fails}}^{\text{unit}})^{E_{1/2/3}(R+1)/2} + k(m_{1/2/3} + 2R)Bp_{\text{f}} \end{aligned}. \tag{8.10}$$

Using the upper approximation turns out to be justified, taking into consideration that in the worst considered case (for $R = 3$ and $P_{\text{fails}}^{\text{unit}} = 20\%$) both addends in (A.8) are smaller than 10% and thus the probability of having both of them failing at the same time is smaller than 1%.

The chip fails if any of the outputs of any reliable block fails; hence, the upper bound probability that the whole chip fails is expressed in (8.11) (when $P_{\text{fails}}^{\text{block}} \ll 1$),

$$\begin{aligned} P_{\text{fails}}^{\text{chip}} &\approx \frac{N}{N_c} F_{\text{out}}(P_{\text{fails}}^{\text{red.unit}} + P_{\text{fails}}^{\text{dec.gate}}) \\ &\approx \frac{N}{N_c} F_{\text{out}}\left(A(4P_{\text{fails}}^{\text{unit}})^{E_{1/2/3}(R+1)/2} + k(m_{1/2/3} + 2R)p_{\text{f}}\right) \end{aligned}. \tag{8.11}$$

Here it is assumed that probabilities of failure of outputs of the reliable block are uncorrelated and that errors in each reliable block are also uncorrelated, i.e., that common-mode (or common cause) failures are not present in the redundant system [247]. This is actually the worst case with respect to reliability.

The optimal partition size ($N_{c,\text{opt}}$, also depicted in Fig. 8.13) for a given p_f is derived as a numerical solution of the expression $dP_{\text{fails}}^{\text{chip}}/dN_c = 0$. After substituting $N_{c,\text{opt}}$ in (8.11), the numerical solution of the equation $dP_{\text{fails}}^{\text{chip}}/dR = 0$ gives the minimal possible value for the probability of the chip failure and redundancy factor that provides it. Substituting R into (8.11) together with the condition$P_{\text{fails}}^{\text{chip}} < 0.1$ reveals the maximal defect density that can be supported in order to achieve a yield $Y > 0.9$. The presented procedure is applied to a chip design ($N = 10^9$). The ratio between two addends in (8.11) is also evaluated.

8.3.1.1 The Chip Design Example

The following assumptions are made:

1. The optimal partition size is larger than 10^5 devices and therefore the partitioning scheme described in Section 8.2.2 is applied and a constant F_{out} is assumed;
2. The exponential factor (E) is assumed constant, $E_1 = E_2 = 1$, $E_3 = 1.9$, and the only benefit of using AVG compared to MV is a relatively smaller realization of the decision gate for AVG compared to MV.
3. The dependence of factor A on N_c can be neglected, since it is much smaller than the dependence of the $(4P_{\text{fails}}^{\text{unit}})^{E_{1/2/3}(R+1)/2}$ factor on N_c.

Taking previous assumptions into consideration and applying the relation $dP_{\text{fails}}^{\text{chip}}/dN_c = 0$ gives the expression for $N_{c,\text{opt}}$:

$$N_{c,\text{opt}} = \frac{dN_c}{dP_{\text{fails}}^{\text{unit}}} P_{\text{fails}}^{\text{unit}} \cdot \frac{2}{E(R+1)} \left(1 + \frac{P_{\text{fails}}^{\text{dec.gate}}}{P_{\text{fails}}^{\text{red.unit}}}\right). \tag{8.12}$$

Equation (8.13) holds around the optimal point:

$$P_{\text{fails}}^{\text{unit}}\Big|_{N_c = N_{c,\text{opt}}} = \frac{dP_{\text{fails}}^{\text{unit}}}{dN_c}\Bigg|_{N_c = N_{c,\text{opt}}} \cdot N_{c,\text{opt}}. \tag{8.13}$$

Combining (8.12) and (8.13) gives

$$\frac{2}{E(R+1)} \left(1 + \frac{P_{\text{fails}}^{\text{dec.gate}}}{P_{\text{fails}}^{\text{red.unit}}}\right)\Bigg|_{N_c = N_{c,\text{opt}}} = 1. \tag{8.14}$$

By rearranging (8.14), the ratio between $P_{\text{fails}}^{\text{red.unit}}$ and $P_{\text{fails}}^{\text{dec.gate}}$ becomes

$$P_{\text{fails}}^{\text{red.unit}} = \frac{1}{\frac{E_{1/2/3}(R+1)}{2} - 1} P_{\text{fails}}^{\text{dec.gate}}. \tag{8.15}$$

Equation (8.15) transforms into

$$P_{\text{fails}}^{\text{unit}}\Big|_{N_c = N_{c,\text{opt}}} = \frac{1}{4} \left(\frac{1}{A\left(\frac{E_{1/2/3}(R+1)}{2} - 1\right)} P_{\text{fails}}^{\text{dec.gate}} \right)^{\frac{2}{E_{1/2/3}(R+1)}}. \tag{8.16}$$

After acquiring the value for $P_{\text{fails}}^{\text{unit}}\big|_{N_c=N_{c,\text{opt}}}$, $N_{c,\text{opt}}$ is directly read from Table 8.9. By combining (8.13) and (8.16), the final expression for $N_{c,\text{opt}}$ can be obtained as:

$$N_{c,\text{opt}} = \frac{1}{4\left.\frac{dP_{\text{fails}}^{\text{unit}}}{dN_c}\right|_{N_c=N_{c,\text{opt}}}} \left(\frac{1}{A\left(\frac{E_{1/2/3}(R+1)}{2} - 1\right)} P_{\text{fails}}^{\text{dec.gate}} \right)^{\frac{2}{E_{1/2/3}(R+1)}}. \quad (8.17)$$

$\frac{dP_{\text{fails}}^{\text{unit}}}{dN_c}$ can be numerically extracted from Table 8.9.

After inserting (8.15) into (8.11) the probability of chip failure for the optimal partition size is

$$P_{\text{fails}}^{\text{chip}}\big|_{N_c=N_{c,\text{opt}}} \approx \frac{N}{N_{c,\text{opt}}} F_{\text{out}} \left(\frac{1}{\frac{E_{1/2/3/}(R+1)}{2} - 1} + 1 \right) P_{\text{fails}}^{\text{dec.gate}}. \quad (8.18)$$

The probability of chip failure considering $R = 3$, a device probability of failure $p_f = 1 \times 10^{-6}$ and different partition sizes is illustrated in Fig. 8.13. N_c in Fig. 8.13 only takes values that yield a maximum probability of unit failure which are smaller than 10% according to assumptions for approximations from Section 8.1. $N_{c,\text{opt}} = 3.2 \times 10^6$ for RMR-MV and RMR-AVG and is significantly higher for RMR-4LRA $N_{c,\text{opt}} = 3.2 \times 10^8$ which is a consequence of a much larger decision gate (six times larger than MV and 11 times larger than AVG). However, the minimum for $P_{\text{fails}}^{\text{chip}}$ is still lower for 4LRA, and the efficiency of 4LRA successfully compensates the drawback of a much larger decision gate.

Expression (8.15) indicates that for the optimal partition size, a dominant cause of chip failure is related to *the unreliability of the decision gate*. The impact of the decision gate on the chip probability of failure is $\frac{E_{1/2/3}(R+1)}{2} - 1$ times larger than the impact of the redundant units. Decision gates can be assumed to be critical elements, which reflects into large values of $N_{c,\text{opt}}$, as a result of a balance between the need to use decision gates, and the necessity to limit their total number.

Accordingly, the probability of chip failure in Fig. 8.13 may be perceived as relatively high for low defect density. However, when compared to a case without redundancy (Table 8.11, where the yield of the analyzed chip is evaluated), the benefit of using the reliable architecture becomes obvious. If only manufacturing defects are considered, then the yield becomes $Y = 1 - P_{\text{fails}}^{\text{chip}}$.

Table 8.11 Yield for chip with 10^9 devices and $p_f = 1 \times 10^{-6}$

Configuration	Yield (%)
No redundancy	0
RMR-MV ($R = 3$)	93.7
RMR-AVG ($R = 3$)	93.9
RMR-4LRA ($R = 3$)	98.3

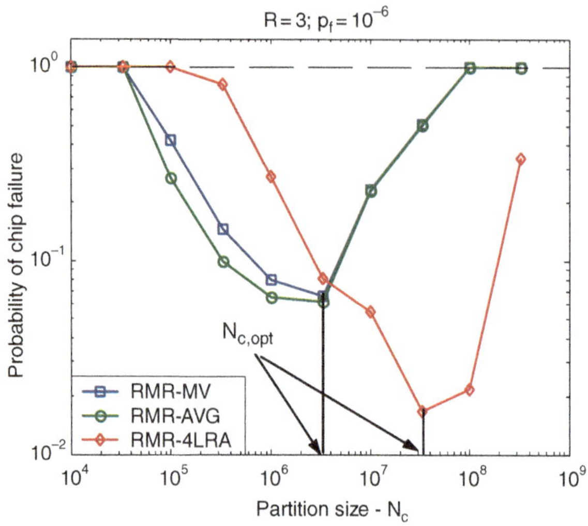

Fig. 8.13 Comparative analysis of RMR-MV, RMR-AVG, and RMR-4LRA in terms of probability of chip failure for different partition sizes ($R = 3$; $p_\mathrm{f} = 1 \times 10^{-6}$)

In Fig. 8.14, the probability of chip failure for the optimal N_c and for selected values of p_f is plotted with respect to the redundancy factor R. Values for p_f are chosen to be equally distributed in the range between the minimal value where redundancy is necessary and the maximal tolerable value (see further Table 8.12). The probability of chip failure is decreasing along with an increase of R. After reaching an optimal point, it increases again, as a result of the decreasing reliability of a decision gate due to a linear increase in the count of its devices (inputs). Regarding decision gates, MV and AVG show similar results. The only difference is due to a smaller realization of AVG compared to MV. 4LRA shows better performance than AVG and MV only for low redundancy factors ($R < 7$) and lower defect densities ($p_\mathrm{f} < 10^{-7}$).

In the following, the maximal defect density that an RMR-based technique can tolerate is determined. Considering that the whole chip actually consists of one single partition ($p_{\mathrm{f,max}}(N = N_\mathrm{c})$) for a maximal defect density and that the desired probability of a chip to operate correctly is over 90%, (8.19) holds (taking (8.18) into consideration):

$$0.1 = F_\mathrm{out}\left(\frac{1}{\frac{E_{1/2/3}(R+1)}{2} - 1} + 1\right) k(m_{1/2/3} + 2R)p_{\mathrm{f,max}} \qquad (8.19)$$

From (8.19), $p_{\mathrm{f,max}}$ for optimal redundancy factor (R_opt), R_opt is 5, 3, and 7 for MV, AVG, and 4LRA decision gates, respectively. After placing the obtained values for R_opt into (8.16) the following values are read from Table 8.9: $N_\mathrm{max} = 1 \times 10^5$, 32, 000, and$1 \times 10^6$ for MV, AVG, and 4LRA decision gates, respectively. Then, the values for F_out are taken from Table 8.7 and placed into (8.19). By solving

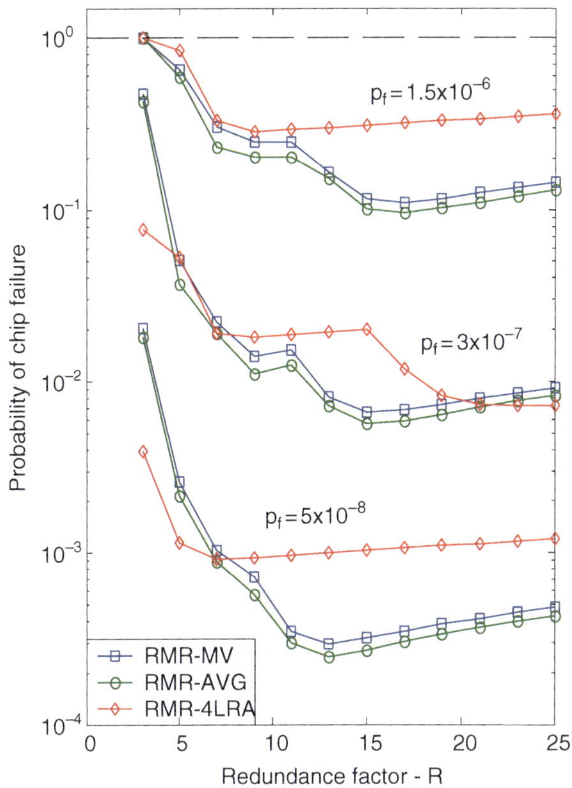

Fig. 8.14 Comparative analysis of RMR-MV, RMR-AVG, and RMR-4LRA in terms of probability of chip failure for different redundancy factors, defect densities, and optimal partition sizes

(8.19) $p_{\mathrm{f,max}} = 8.21 \times 10^{-5}$, 7.7×10^{-5}, and 1.88×10^{-5} for MV, AVG, and 4LRA decision gates, respectively, are obtained.

From (8.18) by setting $P_{\mathrm{fails}}^{\mathrm{chip}} = 0.1$, a maximal size of a chip (maximal effective number of devices) for a given defect density can be derived by numerically solving the equation $\frac{\partial^2 N}{\partial N_{\mathrm{c}} \partial R} = 0$, where N is also derived from (8.18), and given as follows:

$$N = \frac{0.1}{\left(\frac{1}{\frac{E_{1/2/3}(R+1)}{2}-1} + 1\right) k(m_{1/2/3} + 2R) p_{\mathrm{f}}} F_{\mathrm{out}} N_{\mathrm{c}}. \qquad (8.20)$$

The numerical values presented in Table 8.12a–c are obtained applying the presented procedure for different defect densities and for MV, AVG, and 4LRA decision gates, respectively.

Table 8.12 Maximal effective number of devices, optimal redundancy, and partition size values for (a) MV, (b) AVG, and (c) 4LRA decision gates

(a)

p_f	N_{max}	$N_{c,opt}$	R_{opt}
$8.21 \times 10^{-5} = p_{f,max}$	10,000	10,000	5
1×10^{-5}	3.2×10^7	3.2×10^6	15
1.36×10^{-6}	1×10^9	3.2×10^7	17

(b)

p_f	N_{max}	$N_{c,opt}$	R_{opt}
$7.7 \times 10^{-5} = p_{f,max}$	10,000	10,000	3
1×10^{-5}	3.52×10^7	3.2×10^6	15
1.5×10^{-6}	1×10^9	3.2×10^7	15

(c)

p_f	N_{max}	$N_{c,opt}$	R_{opt}
$1.88 \times 10^{-5} = p_{f,max}$	10,000	10,000	7
1×10^{-5}	9.6×10^6	3.2×10^6	9
1.09×10^{-6}	1×10^9	1×10^8	25

Following points are concluded by analyzing Table 8.12:

1. The optimal partition size does not depend on the total chip size, but only on the defect density, redundancy factor, and the type of decision gate.
2. The optimal partition size is reduced for increased defect densities. However, the optimal partition size remains relatively high (in the order of few thousand devices), even in cases of high defect densities ($> 1 \times 10^{-4}$).
3. Considering that optimal partition sizes are very large ($N_c > 10^5$) for all decision gate realizations, the area increase related to decision gates is always small (<5%); thus, it is justified to state that the area overhead $N_{tot}/10^9 = R$, where N_{tot} is the total number of devices on the chip.
4. The maximal defect density that allows building a chip with 10^9 effective devices and with a probability of correct operation over 90% is equal to 1.55×10^{-6} for AVG decision gate. A chip without redundancy with the same constraints could only be built if the defect density was smaller than 5×10^{-10}. Thus, the region of interest for the application of RMR is defined by defect densities interval $p_f \in [5 \times 10^{-10}, 1.55 \times 10^{-6}]$.
5. The redundancy factor has an optimum. For increased defect densities this value is reducing. In the region of interest, the optimal R is never larger than 25.
6. In the region of interest, the optimal partition size is in the order of millions of devices, typically representing a processing unit (ALU, DSP core, etc.) in a modern integrated system.

8.3.2 Cascaded R-Fold Modular Redundancy (CRMR)

An improved fault-tolerant technique must be developed in order to realize 10^9 effective devices on-chip, for defect densities higher than 1.55×10^{-6}. In the literature [6, 114], CRMR has always been considered as CTMR (cascaded triple modular redundancy) and no effective procedure for optimizing the partition size or the redundancy factors have been provided. Therefore, CRMR has been considered to provide little advantage over RMR and no efficient application potentiality. In the following, it is demonstrated that a suitable window of operation can be determined for the CRMR fault-tolerant technique, using an appropriate optimization procedure.

For CRMR, the maximal tolerable defect density remains the same as in the case of RMR ($p_{\mathrm{f,max}} = 5.5 \times 10^{-5}$). However, as demonstrated in the following, operational chips with more effective devices than in the case of RMR could be built, when using optimized CRMR in case of lower defect densities ($<1\times10^{-5}$).

The optimization procedure considering the "first-order" CRMR is derived. This procedure is generalized to CRMR of any order, in a further step. A "first-order" CRMR architecture is depicted in Fig. 8.15. It consists of two layers, namely the "zeroth layer" and the "first layer". In the general case, the "nth-order" CRMR consists of $n+1$ layers. The "zeroth layer" of CRMR actually represents RMR, where N_{c0} devices are grouped into partitions, replicated R_0 times, and a voter is included, forming a reliable module. In the "first layer" of CRMR, N_{c1} of these modules are grouped into a new partition, replicated R_1 times, and a voter is included, thus forming a group (Fig. 8.15). A chip consists of $\frac{N}{N_{\mathrm{c0}}N_{\mathrm{c1}}}$ groups.

Partitions in both layers are assumed to have the same number of outputs (bundle size F_{out}), and outputs of all the partitions are grouped into the common bus (Fig. 8.15). All the assumptions and approximations taken for the RMR chip design example are also valid for CRMR.

For "zeroth layer" of CRMR equation (8.18) can be written as

$$P_{\mathrm{fails,0}}^{\mathrm{chip}} \approx \frac{N}{N_{\mathrm{c0}}} F_{\mathrm{out}} \left(P_{\mathrm{fails,0}}^{\mathrm{red.unit}} + P_{\mathrm{fails,0}}^{\mathrm{dec.gate}} \right) = \frac{N}{N_{\mathrm{c0}}} F_{\mathrm{out}} P_{\mathrm{F0}}, \tag{8.21}$$

where $P_{\mathrm{fails,0}}^{\mathrm{chip}}$ is the probability of failure of a chip, $P_{\mathrm{fails,0}}^{\mathrm{red.unit}}$ is the probability that R_0 redundant units fail, $P_{\mathrm{fails,0}}^{\mathrm{dec.gate}}$ is the probability that the decision gate fails, all of them given for the RMR ("zeroth-order" CRMR) fault-tolerant technique. P_{F0} represents the probability that an output of the reliable block ("zeroth layer") fails and, according to (8.18), is given as (considering $N_{\mathrm{c,opt}}$)

$$P_{\mathrm{F0}} = \left(\frac{1}{\frac{E_{1/2/3/(R_0+1)}}{2} - 1} + 1 \right) P_{\mathrm{fails,0}}^{\mathrm{dec.gate}}. \tag{8.22}$$

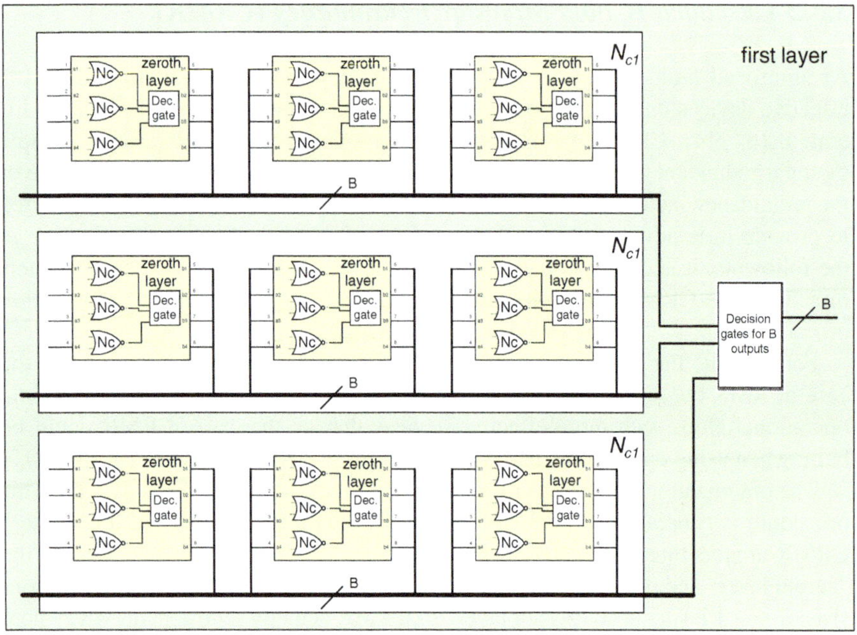

Fig. 8.15 Schematic representation of "first-order" CRMR

Repeating this procedure for the "first-order" CRMR, the probability of chip failure is derived as

$$\begin{aligned} P_{\text{fails},1}^{\text{chip}} &= \frac{N}{N_{\text{c}0}N_{\text{c}1}} F_{\text{out}} \left(P_{\text{fails},1}^{\text{red.u.}} + P_{\text{fails},1}^{\text{voter}} \right) \\ &= \frac{N}{N_{\text{c}0}N_{\text{c}1}} F_{\text{out}} \left(\frac{1}{\frac{E_{1/2/3/(R_1+1)}}{2} - 1} + 1 \right) P_{\text{fails},1}^{\text{dec.gate}}, \end{aligned} \tag{8.23}$$

where $P_{\text{fails},1}^{\text{chip}}$ is the probability of failure of a chip, $P_{\text{fails},1}^{\text{red.unit}}$ is the probability that R_1 redundant units fail, $P_{\text{fails},1}^{\text{dec.gate}}$ is the probability that a decision gate fails, and N_{c1} is the partition size; all given for "first layer" of CRMR. An assumption is that a unit output in "first layer" of CRMR (consisting of N_{c1} reliable blocks of "zeroth layer") fails if a corresponding output of any reliable blocks fails. Following this, it is also valid

$$P_{\text{fails},1}^{\text{unit}} = N_{\text{c}1} P_{\text{F}0}, \tag{8.24}$$

where $P_{\text{F}0}$ represents the probability that an output of the reliable block in the "zeroth layer" fails. $N_{\text{c}1,\text{opt}}$ is derived in the same way as $N_{\text{c},\text{opt}}$ in (8.17)

$$N_{\text{c1,opt}} = \frac{1}{4\left.\frac{dP_{\text{fails,1}}^{\text{unit}}}{dN_{\text{c1}}}\right|_{N_{\text{c1}}=N_{\text{c1,opt}}}} M(R_1)^{\frac{2}{E_{1/2/3}(R+1)}} = \frac{1}{4P_{\text{F0}}} M(R_1)^{\frac{2}{E_{1/2/3}(R+1)}},$$
$$M(R_i) = \left(\frac{1}{A\left(\frac{E_{1/2/3}(R_1+1)}{2} - 1\right)} P_{\text{fails},i}^{\text{dec.gate}} \right). \tag{8.25}$$

The optimal partition size for a "zeroth layer" in CRMR is exactly the same size as the optimal partition size for RMR, as derived in (8.17), i.e., $N_{\text{c0,opt}} = N_{\text{c,opt}}$. The proof is presented in Appendix A.1.

After substituting the value from (8.17) for $N_{\text{c0,opt}}$ into (A.3)

$$P_{\text{fails,1}}^{\text{chip}} = \frac{4^2 N \left.\frac{dP_{\text{fails}}^{\text{unit}}}{dN_{\text{c}}}\right|_{N_{\text{c}}=N_{\text{c0,opt}}}}{M(R_1)M(R_0)} \left(\frac{1}{\frac{E_{1/2/3/}(R_1+1)}{2} - 1} + 1 \right) P_{\text{fails,1}}^{\text{voter}} \cdot \left(\frac{1}{\frac{E_{1/2/3/}(R_0+1)}{2} - 1} + 1 \right) P_{\text{fails,0}}^{\text{dec.gate}} \tag{8.26}$$

Considering the general case of the "nth-order" CRMR, expression (A.12) transforms into

$$P_{\text{fails},n}^{\text{chip}} = f(R_0, R_1, \ldots, R_n) = N \left.\frac{dP_{\text{fails}}^{\text{unit}}}{dN_{\text{c}}}\right|_{N_{\text{c}}=N_{\text{c0,opt}}} \prod_{i=0}^{n} \frac{4P_{\text{F}i}}{M(R_i)}, \tag{8.27}$$

where $i \in [0,\ n-1]$. The details are presented in Appendix A.1.

Minimizing $f(R_0, R_1, \ldots, R_n)$ with respect to R_i, $i \in [0,\ n]$ becomes trivial due to the fact that a probability of chip failure can be split into functions, where each function depends on one single redundancy factor $f(R_0, R_1, \ldots, R_n) = g(R_0) \cdot g(R_1) \cdot \cdots \cdot g(R_n)$. The optimum is achieved for $\partial\left(\frac{P_{\text{F}i}}{M(R_i)}\right)/\partial R_i = 0$, for $i \in [0,\ n]$. The optimum is given as

$$R_{0,\text{opt}} = R_{1,\text{opt}} = \cdots = R_{n,\text{opt}} = R_{\text{opt}}. \tag{8.28}$$

Furthermore, at the optimum, the probability of "ith-layer" failure ($i \in [0,\ n]$) is given as

$$P_{\text{F0}} = P_{\text{F1}} = \cdots = P_{\text{F}n} = P_{\text{F,min}}. \tag{8.29}$$

The optimal partition size for "ith layer," $i \in [1,\ n]$ is given as

$$N_{\text{c1,opt}} = N_{\text{c2,opt}} = \cdots = N_{\text{c}n\text{,opt}}. \tag{8.30}$$

Defining the *gain* as the ratio of the probability of chip failure between CRMR of "jth order" and "$(j+1)$th order" (between "jth and $(j+1)$th layer") yields

$$\text{gain} = \frac{P_{\text{fails},j}^{\text{chip}}}{P_{\text{fails},j+1}^{\text{chip}}} = \frac{M(R_{j+1})}{2P_{\text{F}j+1}} = \frac{M(R_{opt})}{2P_{\text{F,min}}} = N_{\text{c1,opt}}.$$

In a special case, the *gain* between RMR and the "first-order" CRMR is also equal $N_{\text{c1,opt}}$. Results considering this special case, for a maximal effective number of devices, an optimal redundancy factor and *gain* are shown in Table 8.13a–c for different defect densities and for MV, AVG, and 4LRA decision gates, respectively. The analysis of Table 8.13 reveals that CRMR becomes ineffective for defect densities higher than 1×10^5, because the increase in the necessary redundancy becomes higher than the *gain*. The *gain* between the no-redundancy case and RMR can also be expressed as $M(R_0)/2P_{\text{F0}}$.

Table 8.13 Maximal effective number of devices, optimal redundancy, and partition size values in case of RMR and CRMR for (a) MV, (b) AVG, and (c) 4LRA decision gates

(a)

p_f	$N_{\text{max,CRMR}}$	$N_{\text{max,RMR}}$	$N_{\text{c0,opt}}$	$N_{\text{c1,opt}}$ **=gain**	$R_{0,\text{opt}} = R_{1,\text{opt}}$
8.21×10^{-5}	10,000	10,000	10,000	**1**	5
1×10^{-5}	2.19×10^{10}	3.2×10^{7}	3.2×10^{6}	**685**	15
1.36×10^{-6}	4.52×10^{12}	1×10^{9}	3.2×10^{7}	**4519**	17

(b)

p_f	$N_{\text{max,CRMR}}$	$N_{\text{max,RMR}}$	$N_{\text{c0,opt}}$	$N_{\text{c1,opt}}$ **= gain**	$R_{0,\text{opt}} = R_{1,\text{opt}}$
7.7×10^{-5}	10,000	10,000	10,000	**1**	3
1×10^{-5}	2.36×10^{10}	3.52×10^{7}	3.2×10^{6}	**671**	15
1.5×10^{-6}	4.31×10^{12}	1×10^{9}	3.2×10^{7}	**4031**	17

(c)

p_f	$N_{\text{max,CRMR}}$	$N_{\text{max,RMR}}$	$N_{\text{c0,opt}}$	$N_{\text{c1,opt}}$ **= gain**	$R_{0,\text{opt}} = R_{1,\text{opt}}$
1.88×10^{-5}	10,000	10,000	10,000	**1**	7
1×10^{-5}	5.96×10^{9}	9.6×10^{6}	3.2×10^{6}	**621**	21
1.09×10^{-6}	5.71×10^{12}	1×10^{9}	1×10^{8}	**5707**	25

For CRMR, a total redundancy factor can be defined as a product of redundancy factors of each layer in "nth-order CRMR," $R_{\text{tot}} = R_{1,\text{opt}} R_{1,\text{opt}} \cdots R_{1,\text{opt}} = (R_{\text{opt}})^n$. The maximal tolerable defect density, the total redundancy factor, and gain between layers have been determined for RMR and CRMR of different orders for (a) MV, (b) AVG, and (c) 4LRA decision gates and are presented in Table 8.14. With respect

Table 8.14 Maximal tolerable defect density, total redundancy factor, and *gain* in case of RMR and CRMR for (a) MV, (b) AVG, and (c) 4LRA decision gates

(a)

	$p_{\mathrm{f,max}}$	R_{tot}	Gain
Third order CRMR	1.88×10^{-4}	28,561	41
Second order CRMR	8.51×10^{5}	4,913	72
First order CRMR	2.5×10^{-5}	225	274
RMR	1.36×10^{-6}	17	4,519

(b)

	$p_{\mathrm{f,max}}$	R_{tot}	Gain
Third order CRMR	1.91×10^{-4}	28,561	40
Second order CRMR	8.51×10^{-5}	4,913	72
First order CRMR	2.65×10^{-5}	225	260
RMR	1.5×10^{-6}	17	4,031

(c)

	$p_{\mathrm{f,max}}$	R_{tot}	Gain
Third order CRMR	1.45×10^{-4}	2,401	53
Second order CRMR	7.36×10^{-5}	729	84
First order CRMR	1.76×10^{-5}	169	272
RMR	1.09×10^{-6}	25	5,707

to this table, considering an exponential increase of the necessary redundancy and a reduction in the *gain* when increasing the order of CRMR, only the "first-order" and the "second-order" CRMR architectures can have practical applications. The maximal tolerable defect density is more than 15× higher using a "first-order" CRMR, and 63× higher using a "second-order" CRMR, compared to RMR. The improvement in the maximal tolerable defect density from "second-order" to "third-order" CRMR is only 2.2× and is not sufficient to justify the use of "third-order" CRMR. Considering the different realizations, MV and AVG show almost the same performance regarding the maximal tolerable defect density, while 4LRA is worse for all orders of CRMR. However, 4LRA needs a significantly lower total redundancy factor than MV and AVG.

8.3.3 Distributed R*-Fold Modular Redundancy (DRMR)*

The chip consisting of RMR reliable blocks fails if redundant units and/or the decision gate of at least one reliable block fails. Therefore, the input signals of the reliable block are always assumed to be error free. The chip consisting of DRMR-reliable blocks on the other hand can operate correctly even when some of the decision gates are failing. However, with some failing decision gates, the input signals of the reliable blocks can be erroneous in specific cases and they also have to be taken into consideration. Hence, when using the DRMR fault-tolerant technique, the probability of an output of the reliable block to fail ($P_{\text{fails}}^{\text{block}}$) depends on the probability of failure of redundant units ($P_{\text{fails}}^{\text{red.unit}}$) and the probability of failure of input signals ($P_{\text{fails}}^{\text{in.sig.}}$). Considering that $P_{\text{fails}}^{\text{block}}$ depends on $P_{\text{fails}}^{\text{in.sig.}}$, even though the errors in each block are assumed to be uncorrelated, there would be a correlation between input signals coming into different blocks due to the fanout of the block output signals. This correlation could also be neglected, which will be shown further.

The failure of the unit input signals causes a unit to fail. Here, it is assumed that a unit fails if only one of the input signals of that unit fails, e.g., the input signal is either a logic-0 when its expected value is logic-1, or conversely. The probability of the single input signal failing is equal to the probability of failure of the decision gate driving this signal.

Following discussions in Section 8.1 where group of R redundant units fails when at least $(R+1)/2$ units fail, the probability of failure of R redundant units out of which i units have already failed due to erroneous input signals ($P_{\text{fails},(i)}^{\text{red.unit}}$) is given as

$$P_{\text{fails},(i)}^{\text{red.unit}} = 2\binom{R-i}{\frac{R+1}{2}-i}(1-P_{\text{fails}}^{\text{unit}})^{(R-3)/2} \cdot \left(1-\frac{4-2i}{R+3-2i}P_{\text{fails}}^{\text{unit}}\right)(P_{\text{fails}}^{\text{unit}})^{E_{1/2/3}\left(\frac{R+1}{2}-i\right)}, \tag{8.31}$$

where indices 1, 2, and 3 of the exponential coefficient $E_{1/2/3}$ stand for MV, AVG, and 4LRA decision gates, respectively. Equation (8.31) is valid for $i \in [1, \frac{R-1}{2}]$. For $i \geq \frac{R+1}{2}$, $P_{\text{fails},(i)}^{\text{red.unit}} = 1$. It is assumed here that all three types of decision gates can be used. However, the optimal size of the partition in DRMR technique is too small for an efficient application of 4LRA. This will be shown further in this section. Moreover, because of the smaller partition size, $P_{\text{fails}}^{\text{unit}} < 1\%$ is valid and in all possible applications of DRMR, the $\left(1-\frac{4-2i}{R+3-2i}P_{\text{fails}}^{\text{unit}}\right)$ factor can be omitted from (8.31).

The probability that input signals to i out of R redundant units are erroneous ($P_{\text{fails},(i)}^{\text{in.sig.}}$) can be expressed as

$$P_{\text{fails},(i)}^{\text{in.sig.}} = \binom{R}{i}\left[F_{\text{in}}P_{\text{fails}}^{\text{dec.gate}}\right]^i(1-P_{\text{fails}}^{\text{dec.gate}})^{(R-i)F_{\text{in}}}, \tag{8.32}$$

The calculation details are presented in Appendix A.2.

The probability of an output of the reliable block to fail when n units fail due to input signals ($P^{\text{block}}_{\text{fails},(i)}$) is given as

$$\begin{aligned} P^{\text{block}}_{\text{fails},(i)} &= P^{\text{red.unit}}_{\text{fails},(i)} \cdot P^{\text{in.sig.}}_{\text{fails},(i)} \\ &= 2\binom{R-i}{\frac{R+1}{2}-i}\left(1-P^{\text{unit}}_{\text{fails}}\right)^{(R-3)/2}\left(P^{\text{unit}}_{\text{fails}}\right)^{E_{1/2/3}\left(\frac{R+1}{2}-i\right)} \\ &\cdot \binom{R}{i}\left(F_{\text{in}}P^{\text{dec.gate}}_{\text{fails}}\right)^{i}\left(1-P^{\text{dec.gate}}_{\text{fails}}\right)^{(R-i)F_{\text{in}}} \\ &= 2\binom{R}{\frac{R+1}{2}}\binom{\frac{R+1}{2}}{i}\left(1-P^{\text{unit}}_{\text{fails}}\right)^{(R-3)/2}\left(1-P^{\text{dec.gate}}_{\text{fails}}\right)^{\frac{R-1}{2}F_{\text{in}}} \\ &\cdot \left(P^{\text{unit}}_{\text{fails}}\right)^{E_{1/2/3}\left(\frac{R+1}{2}-i\right)}\left(1-P^{\text{dec.gate}}_{\text{fails}}\right)^{\left(\frac{R+1}{2}-i\right)F_{\text{in}}}\left(F_{\text{in}}P^{\text{dec.gate}}_{\text{fails}}\right)^{i} \end{aligned} \quad . \quad (8.33)$$

The equality $\binom{R-i}{\frac{R+1}{2}-i}\binom{R}{i} = \binom{R}{\frac{R+1}{2}}\binom{\frac{R+1}{2}}{i}$ is used in (8.33).

The total probability for an output of the reliable block to fail is equal to the sum of probabilities of reliable block failure when i out of R redundant units fail due to input signals for $i \in [0,\ R]$:

$$\begin{aligned} P^{\text{block}}_{\text{fails}} &= \sum_{i=0}^{\frac{R-1}{2}} P^{\text{red.unit}}_{\text{fails},(i)} \cdot P^{\text{in.sig.}}_{\text{fails},(i)} + P^{\text{in.sig.}}_{\text{fails},(\frac{R+1}{2})} + \sum_{i=\frac{R+3}{2}}^{R} P^{\text{in.sig.}}_{\text{fails},(i)} \\ &\approx A_{\text{DRMR}}\left[(P^{\text{unit}}_{\text{fails}})^{E_{1/2/3}}(1-P^{\text{dec.gate}}_{\text{fails}}) + F_{\text{in}}P^{\text{dec.gate}}_{\text{fails}}\right]^{\frac{R+1}{2}} + \sum_{i=\frac{R+3}{2}}^{R} P^{\text{in.sig.}}_{\text{fails},(i)} \end{aligned}, \quad (8.34)$$

where

$$A_{\text{DRMR}} = 2\binom{R}{\frac{R+1}{2}}(1-P^{\text{unit}}_{\text{fails}})^{(R-3)/2} \cdot (1-P^{\text{dec.gate}}_{\text{fails}})^{\frac{R-1}{2}F_{\text{in}}}.$$

The factor $(1-P^{\text{dec.gate}}_{\text{fails}})$ can be further omitted from (8.34) since $P^{\text{dec.gate}}_{\text{fails}} < 1\%$. For the same reason the following expression is also valid:

$$\begin{aligned} &A_{\text{DRMR}}\left[(P^{\text{unit}}_{\text{fails}})^{E_{1/2/3}} + F_{\text{in}}P^{\text{dec.gate}}_{\text{fails}}\right]^{\frac{R+1}{2}} \gg \\ &\gg \sum_{i=\frac{R+3}{2}}^{R} P^{\text{in.sig.}}_{\text{fails},(i)} \approx \binom{R}{\frac{R+1}{2}}(F_{\text{in}}P^{\text{dec.gate}}_{\text{fails}})^{\frac{R+1}{2}}, \end{aligned} \quad (8.35)$$

and the final expression for the probability of an output of the reliable block to fail is expressed as

$$P_{\text{fails}}^{\text{block}} = A_{\text{DRMR}} \left[(P_{\text{fails}}^{\text{unit}})^{E_{1/2/3}} + F_{\text{in}} P_{\text{fails}}^{\text{dec.gate}} \right]^{\frac{R+1}{2}}. \tag{8.36}$$

The joint probability of failure for two or more reliable blocks is negligible compared to $P_{\text{fails}}^{\text{block}}$. This will be demonstrated by calculating the upper bound of joint probability of failure for two reliable blocks ($P_{\text{fails},(2)}^{\text{block}}$) in the extreme case when two blocks share all inputs. This is the absolute limit since in reality, two blocks usually share only a small fraction of their inputs. Similar to (8.34)

$$P_{\text{fails},(2)}^{\text{block}} = \sum_{i=0}^{\frac{R-1}{2}} \sum_{j=0}^{\frac{R-1}{2}} P_{\text{fails},(i)}^{\text{red.unit}} \cdot P_{\text{fails},(j)}^{\text{red.unit}} \cdot P_{\text{fails},(ij)}^{\text{in.sig.}} + \sum_{i=\frac{R+1}{2}}^{R} \sum_{j=\frac{R+1}{2}}^{R} P_{\text{fails},(ij)}^{\text{in.sig.}},$$

where

$$P_{\text{fails},(ij)}^{\text{in.sig.}} = \begin{cases} P_{\text{fails},(i)}^{\text{in.sig.}} & \text{for } i = j \\ 0 & \text{for } i \neq j \end{cases}. \tag{8.37}$$

Equation (8.37) further transforms into

$$P_{\text{fails},(2)}^{\text{block}} = \sum_{i=0}^{\frac{R-1}{2}} \left(P_{\text{fails},(i)}^{\text{red.unit}} \right)^2 \cdot P_{\text{fails},(i)}^{\text{in.sig.}} + \sum_{i=\frac{R+1}{2}}^{R} P_{\text{fails},(i)}^{\text{in.sig.}} \approx \sum_{i=\frac{R+1}{2}}^{R} P_{\text{fails},(i)}^{\text{in.sig.}}, \tag{8.38}$$

since the ratio between "$(i-1)$th" and "ith" elements of the sum $\sum_{i=0}^{\frac{R-1}{2}} \left(P_{\text{fails},(i)}^{\text{red.unit}} \right)^2 \cdot P_{\text{fails},(i)}^{\text{in.sig.}}$ is

$$\frac{i}{R-i} \frac{(P_{\text{fails}}^{\text{unit}})^{2E_{1/2/3}}}{F_{\text{in}} P_{\text{fails}}^{\text{dec.gate}}} \ll 1. \tag{8.39}$$

and is also valid:

$$\left(P_{\text{fails},(\frac{R-1}{2})}^{\text{red.unit}} \right)^2 P_{\text{fails},(\frac{R-1}{2})}^{\text{in.sig.}} \ll \sum_{i=\frac{E+1}{2}}^{R} P_{\text{fails},(i)}^{\text{in.sig.}}.$$

Finally, taking (8.35) into consideration, it becomes clear that $P_{\text{fails},(2)}^{\text{block}}$ from (8.38) is negligible compared to $P_{\text{fails}}^{\text{block}}$ from (8.36). Since the joint probabilities of reliable blocks are negligible, the chip fails if any of the outputs of any reliable block fails; hence, the upper bound probability that the whole chip fails is

$$P_{\text{fails}}^{\text{chip}} \approx \frac{N}{N_c} F_{\text{out}} A_{\text{DRMR}} \left[(P_{\text{fails}}^{\text{unit}})^{E_{1/2/3}} + F_{\text{in}} P_{\text{fails}}^{\text{dec.gate}} \right]^{\frac{R+1}{2}}. \tag{8.40}$$

Here, similarly as for RMR and CRMR, it is assumed that probabilities of failure of outputs of the reliable block are uncorrelated and that errors in each reliable block are uncorrelated, i.e., that common-mode (or common cause) failures are not present in the redundant system [247], which is actually the worst case. The values for F_{in}, F_{out}, $E_{2/3}$, and $P_{\text{fails}}^{\text{unit}}$ used in calculations of $P_{\text{fails}}^{\text{chip}}$ are taken from Tables 8.7, 8.10, and 8.9, respectively.

Since $dP_{\text{fails}}^{\text{chip}}/dN_c < 0$ is valid for any positive integer value of N_c and that $dP_{\text{fails}}^{\text{chip}}/dR > 0$ is also valid for any integer value of $R \geq 3$, the probability of chip failure always reduces with the decrease of N_c and increase of R. This is illustrated in Fig. 8.16 where the probability of chip failure is given for a defect density $p_f = 5 \times 10^{-6}$, different partition sizes, and redundancy factors for the MV decision gate.

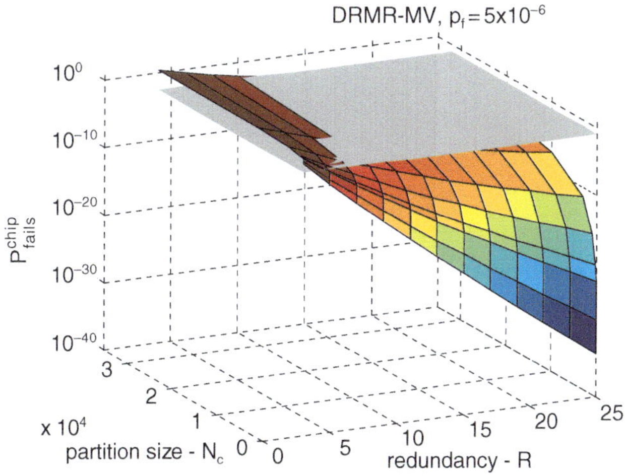

Fig. 8.16 The probability of chip failure for different partition sizes and redundancy factors for the MV decision gate and the reliability constraint threshold surface ($p_f = 5 \times 10^{-6}$)

On the other hand, the total number of devices (N_{tot}) after applying the fault-tolerant architecture can be expressed as

$$N_{\text{tot}} = NR\left(1 + F_{\text{out}}\frac{m_{1/2/3} + 2R}{N_c}\right), \tag{8.41}$$

since one decision gate corresponds to each output of each reliable block. The ratio between the total and the effective number of devices $\left(\frac{N_{\text{tot}}}{N} = \frac{N_{\text{tot}}}{10^9}\right)$ is referred to as the total overhead. According to (8.41) the overhead always increases with the decrease of N_c and increase of R. This is illustrated in Fig. 8.17 where N_{tot} is given for different partition sizes and redundancy factors.

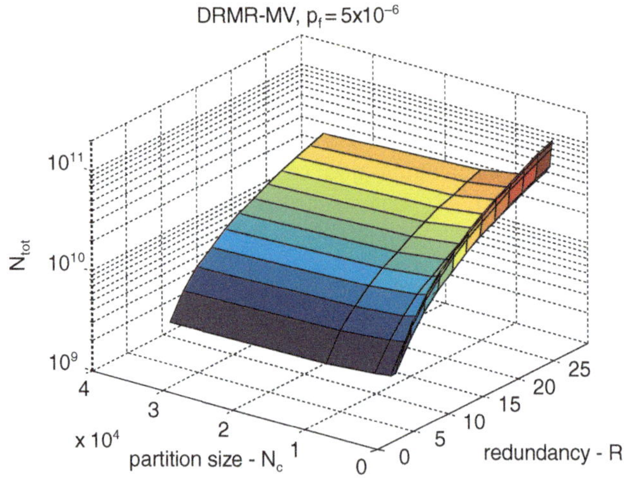

Fig. 8.17 Total number of devices for different partition sizes and redundancy factors and for the MV decision gate

Therefore, the goal of multioptimization can be expressed as

The optimization goal is to minimize the overhead (total number of devices), at the same time guarantee that the probability of chip failure remains smaller than the given constraint.

The design constraint to achieve a yield $Y > 0.9$ gives the condition $P_{\text{fails}}^{\text{chip}} < 0.1$. This condition (also illustrated in Fig. 8.16) applied to (8.40) defines the set of acceptable pairs in the two-dimensional (N_c, R) space. The space of acceptable pairs is depicted in Fig. 8.18. The minimum value of N_{tot} for (N_c, R) pairs belonging to the defined space yields the searched optimum given in Fig. 8.19. Numerical values for the optimum N_c, R, and total overhead $N_{\text{tot}}/10^9$ are given in Table 8.15 for MV and AVG decision gates and for four values of defect densities ranging from moderate to maximum tolerable (determined by the excessive overhead). Values for 4LRA are not given since there is no pair (N_c, R) which satisfies the condition $P_{\text{fails}}^{\text{chip}} < 0.1$. The reason for this is a very large realization of the 4LRA decision gate compared to the partition size.

Even when defect density is much higher than the maximum-tolerable defect density for RMR ($p_f = 1.5 \times 10^{-6}$), the optimization goal has been achieved with relatively small overhead (see Table 8.15). The optimal size of the partition is few orders of magnitude smaller than the values for RMR and CRMR. Comparison of results for MV and AVG decision gates shows that the overhead is in average 30% smaller when AVG is used.

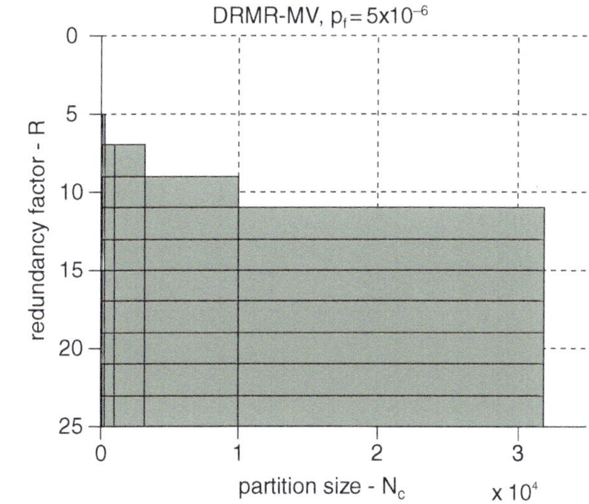

Fig. 8.18 The space of possible values of partition size and redundancy that satisfy the reliability constraint

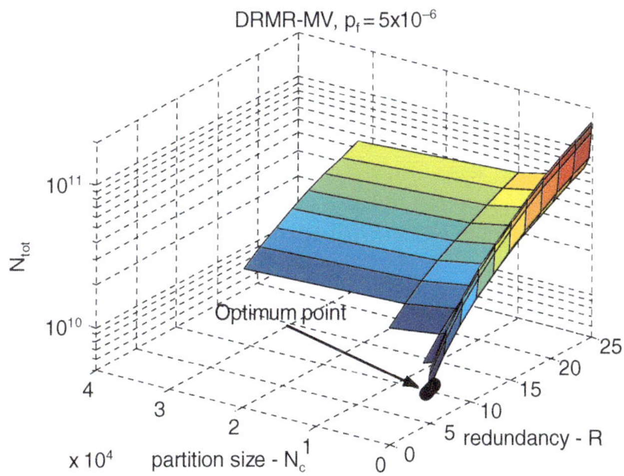

Fig. 8.19 Total number of devices for values of partition size and redundancy that satisfy the reliability constraint and optimal point

8.3.4 NAND Multiplexing

The reliability of NAND multiplexing realized as a parallel restitution [123] is also assessed. The procedure described in [123] is applied in the following for this purpose. Several approaches can be considered in the reliability evaluation of NAND multiplexing. Combinatorial analytical reliability models are very

Table 8.15 Optimal partition size, redundancy, and total overhead for three defect densities and MV and AVG decision gates

p_f	Decision gate	$N_{c,opt}$	R_{opt}	$N_{tot}/10^9$
5×10^{-6}	DRMR-MV	320	5	8.28
	DRMR-AVG	1,000	5	7.52
5×10^{-5}	DRMR-MV	32	7	18.81
	DRMR-AVG	320	9	18.28
3×10^{-4}	DRMR-MV	10	11	39.6
	DRMR-AVG	10	9	28.8
2×10^{-3}	DRMR-MV	10	35	294
	DRMR-AVG	10	27	183.6

accurate [13, 121, 122, 126], but are usually extremely time-consuming in a software implementation. Models that assume normal error distribution [13, 55] are fast with respect to software implementation, but very inaccurate when considering redundancy factors smaller than 1,000. In this section, the model proposed by Han and Jonker [55] which represents a compromise between speed and accuracy is used. The model evaluates the error distribution for the NAND-multiplexing technique by examining each NAND gate in the executive stage, independently.

A binomial distribution describes the number of asserted outputs from the executive unit, and a Markov chain models the output distribution after multiple stages. Since the stages of the NAND multiplexer are represented as a Markov process [55], the output distribution of any stage of the system is only dependent on the distribution of the previous stage. Thus, it is possible to derive a matrix of conditional probabilities of one-stage output signals as a function of the stage input signals. This matrix along with the initial input distribution determines the output distribution of any stage.

This chip is built only using NAND gates as an executive stage which is always followed by two restorative stages (Fig. 8.20).

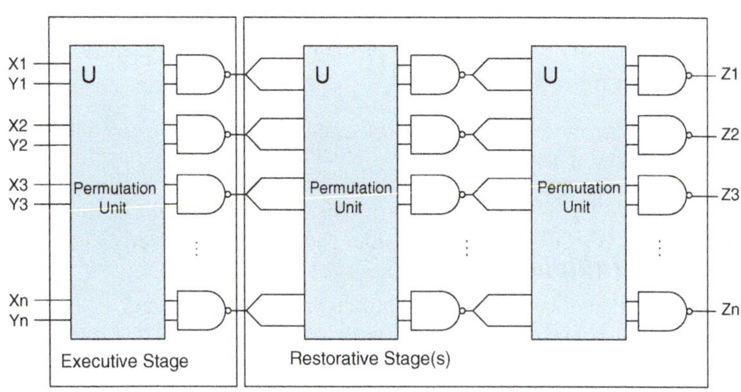

Fig. 8.20 NAND multiplexer

To determine the necessary redundancy factor, a logic depth L is assumed. The probability of any number of outputs of a NAND-multiplexing stage of being incorrect is reduced with the increase in the number of stages [13, 55]. Therefore, the worst-case scenario is represented with the minimal logic depth. The logic depth of $L = 10$ is assumed as the worst case. The model consisting of one chain of NAND multiplexers of a logic depth L is depicted in Fig. 8.21. The probability of ζ outputs after each "ith" stage being stimulated (equal to logic-1) is given as $P(\zeta_i)$. The total number of stages in the chain is $3L = 30$ and each gate is represented with one executive and two restorative stages. Fault-free inputs are assumed, as well as less than 10% of faulty outputs at the end of a chain. Assuming that the correct value of output is stimulated and that bundle size is B, the result is evaluated as correct only if $0.9B$ outputs are stimulated. This probability is denoted as $P_{\text{fails}}^{\text{chain}} = \Pr(\text{num. stimul. outputs at chain end} < 0.9B)$.

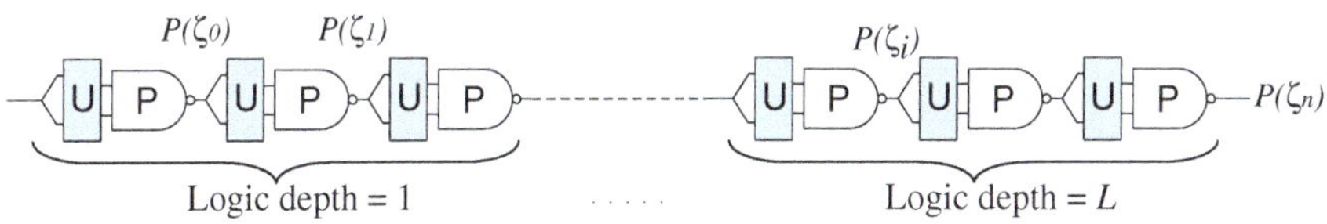

Fig. 8.21 Model of a NAND multiplexer chain of a logic depth L

Each NAND gate is assumed to be built out of four devices and according to (8.1) (Section 8.1) $P_{\text{fails}}^{\text{NAND}} = 4kp_{\text{f}} = 0.8p_{\text{f}}$.

The effective number of devices (only devices in the executive stages) in a chain is $4L = 40$ since each NAND gate is built of four devices. The chip consisting of 10^9 effective devices has 2.5×10^7 chains. The probability of chain failure $P_{\text{fails}}^{\text{chain}}$ has to be lower than 4×10^{-9} since the probability of chip failure is kept lower than 10%. For each defect density, the minimal necessary redundancy factor is determined, which causes a probability of chain failure lower than 4×10^{-9}, using the model described in [55].

8.3.5 Chip-Level Analysis

In the following, the goal consisting of building a chip consisting of 10^{12} devices in total, i.e., 10^9 effective devices, where moderate redundancy factors ($R < 1,000$) are considered, is consistently pursued. The assumption that a chip should work with 90% probability is taken into account.

The maximal tolerable defect density (p_{f}) is calculated in each case for the optimal redundancy factor (R), using $P_{\text{fails}}^{\text{chip}} = 0.1$ and $N = 10^9$, and applying expressions (8.18), (A.16), and (8.40) for RMR, CRMR, and DRMR, respectively. The decision gate that has been used in the RMR, CRMR, and DRMR technique is AVG, considering that it shows the best performance compared to MV and 4LRA.

The maximal tolerable defect densities, considering different redundancy factors, for RMR, CRMR of "first order" and "second order," DRMR and NAND-multiplexing fault-tolerant techniques, are shown in Fig. 8.22.

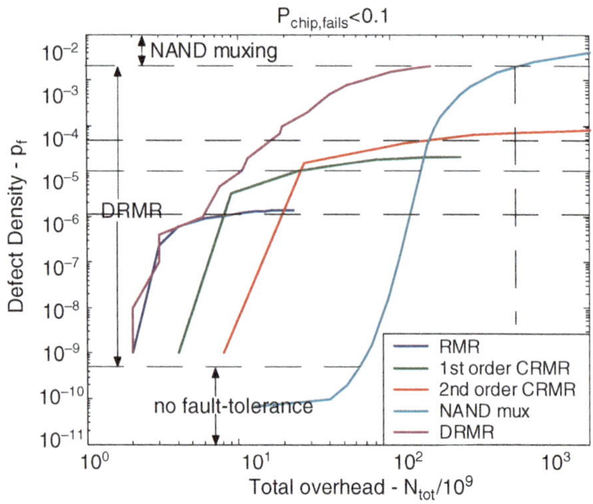

Fig. 8.22 Allowable defect density per device p_f, as a function of the amount of redundancy, R, on a chip, with $N = 10^9$ effective devices, and that must operate correctly with 90% probability, using different fault-tolerant techniques. The window of optimal operation of each technique is also displayed

The whole design space according to defect density can be divided into optimal ranges with respect to fault-tolerant techniques. The upper limit of application of a fault-tolerant scheme is defined as the point where another technique supports higher defect density, at constant overhead. The iterative application of this definition yields three sectors. Each fault-tolerant technique is optimally used in one sector:

1. No redundancy, for defect densities lower than 5×10^{-10}
2. DRMR, for defect densities between 5×10^{-10} and 2×10^{-3}. The total overhead ranges from 2 to 200, being smaller than 10 for all defect densities smaller than 1×10^{-5}
3. NAND multiplexing, for defect densities between 2×10^{-3} and 1×10^{-2}. The minimal total redundancy factor is 595 and 1005. The upper limit has been chosen in accordance with the initial assumption that only moderate overhead as feasible (<1,000).

When DRMR is not available or is not used, the corresponding interval can be divided into three sectors:

1. RMR, for defect densities between 5×10^{-10} and 1×10^{-6}. The minimal redundancy factor is 3 and is sufficient for all defect densities between 5×10^{-10}

and 2×10^{-7}. For defect densities between 2×10^{-7} and 1×10^{-6}, the optimal redundancy factor is increased up to 9

2. "First-order CRMR," for defect densities between 1×10^{-6} and 1×10^{-5}. The minimal total overhead is 9 and is sufficient for all defect densities between 1×10^{-6} and 3×10^{-6}. For defect densities between 3×10^{-6} and 1×10^{-5}, the optimal overhead is 25
3. "Second-order CRMR," for defect densities between 1×10^{-5} and 4.8×10^{-5}. The minimal total overhead is 27 and is sufficient for all defect densities between 1×10^{-5} and 1.8×10^{-5}. For defect densities between 1.8×10^{-5} and 4.8×10^{-5}, the optimal total redundancy factor is increased up to 343.

Observing the presented results, the effectiveness of DRMR should be noted. DRMR represents a universal fault-tolerant technique which can be used in an extremely wide range of defect densities which makes it an excellent choice as the default reliability architecture in the fault-tolerant methodology presented in the beginning of this chapter. Moreover, the averaging gate with fixed optimal threshold (AVG) should be used as a decision gate in our design methodology since it offers the best tradeoff in performance.

8.4 Conclusions

In this chapter the full fault-tolerant methodology is presented and building blocks of the proposed reliability evaluation and optimization procedure are investigated in detail. Important results and conclusions are summarized in the following:

- An estimation of the probability of failure of small to mid-sized circuits has been given with respect to the logic depth of these circuits. The evaluation of the effectiveness of analog averaging techniques in terms of reliability improvement compared to majority voter-based techniques has been performed and it has been proven that the effectiveness significantly reduces with increased logic depth. Moreover, it has been shown that the local optimization can be efficiently performed by resynthesizing circuits, having the logic depth as the minimization goal.
- A fast algorithm for reliability-optimal partitioning has been proposed and realized. The algorithm thrives to solve the problem of combinational circuit partitioning for probability of failure minimization subject to area constraints.
- To the best of our knowledge, the full hierarchical analysis and optimization of hypothetical large-scale (trillion devices) system, from local to system level, has been performed for the first time. Reliability-optimal redundancy factor and partition size as global parameters have been extracted in an analytical way.
- The optimal window of application of each fault-tolerant technique with respect to defect density has been derived. The analysis enables an optimum design trade-off between the reliability and power/area. R-fold modular redundancy with distributed voting and averaging voter has been selected as the most promising

candidate for the implementation in trillion-transistor logic systems. Moreover, it has been demonstrated that target reliability can be achieved with low to moderate redundancy factors ($R < 50$) even for high defect densities (device failure rate up to 10^{-3}).

- A new fault-tolerant design methodology that integrates local-level evaluation, optimal reliability partitioning, and system-level reliability evaluation has been proposed. The methodology is envisioned as an "upgrade" of the existing digital design methodology and new concepts are merged into the methodology in a seamless and transparent way.

The simplistic, nevertheless, realistic hypothesis that the redundancy factor has a monotonic dependence on the area and power dissipation enables creating a link between power/area and reliability (which appears in the above-mentioned discussion as a defect density under which a chip operates with 90% probability). Consequently, the presented method enables selecting the most appropriate fault-tolerant technique, for an optimal power/area budget, and under various defect densities, ranging up to massive defect densities.

Chapter 9
Summary and Conclusions

The microelectronics community has been aware of reliability issues in very deep-submicron and future nanoelectronic technologies for more than a decade and has responded in two major ways. On the one hand, reliability-hardened fabrication technologies have emerged, involving new materials in the fabrication process and new fabrication techniques. Industry has been able to manufacture and commercialize systems of growing complexity following Moore's law. In a parallel track, research mostly carried out in academic institutions has focused on novel methodologies enforcing the principle of fabricating robust systems by construction, tackling the reliability issue at higher level of the semiconductor fabrication flow abstraction hierarchy. While the latter approach currently provides a minor contribution to industrial developments, it can be expected to see its importance grow as the scaling process is observed to exhaust material and technology-based solutions.

This book examines important issues related to the reliable design of nanoscale circuits and systems and aims at providing solutions and research directions to solving these issues at the circuit and systems, architectural, and methodology levels. These solutions include fault-tolerant architectures, fault modeling, reliability evaluation, and design methodology.

9.1 Reliability-Aware Design Methodology

This book presents a wide overview of reliability in VLSI design, the related issues to be tackled, and solutions which have been proposed. Concepts of reliability, historically important and state-of-the-art fault-tolerant techniques and tools for reliability analysis, as well as results of the latest research in nanodevices are methodically introduced and studied. An original approach to reliability-aware design of digital systems is suggested. Original contributions to the state of the art are presented at various levels and are summarized in the following.

A new transistor fault model suitable for CMOS designs is provided to answer the question of inaccurate existing fault models, mainly stuck-at faults, which can be used within reliability evaluation tools. In the model, physical defects are translated

M. Stanisavljević et al., *Reliability of Nanoscale Circuits and Systems*,
DOI 10.1007/978-1-4419-6217-1_9, © Springer Science+Business Media, LLC 2011

into equivalent electrical linear devices such as resistors, capacitors, and nonlinear devices such as diodes and scaled transistors. A total of sixteen possible defects are considered for each transistor. A software tool based on Monte Carlo simulations, incorporating the developed transistor fault model is used for the purpose of reliability analysis of different fault-tolerant architectures at the gate and extended gate levels.

A novel four-layer fault-tolerant hardware architecture (4LRA) is proposed, which is based on a four-layer feed-forward topology. The architecture uses averaging and thresholding as the core voter mechanisms can have fixed or adaptable threshold and is applicable at the gate or extended gate level. It can be applied hierarchically, in a bottom-up way, and combined with other high-level fault absorption techniques. 4LRA has been assessed in terms of reliability and compared to existing fault-tolerant architectures. The comparison has been performed at gate and extended gate levels and different realizations of 4LRA have been considered, namely averaging with fixed threshold (AVG), averaging with adjustable threshold (AVG-opt), and averaging with adaptable threshold (4LRA). Single-ended and differential signaling have been considered, as well as fault-free and faulty voting circuits. All realizations of 4LRA show superior performance compared to R-fold modular redundancy in all evaluations. 4LRA can be applied to circuits built of SET devices, as a typical representation of nanodevice currently under development. A specific fault model suitable for SET devices has been implemented and the Monte Carlo tool has been enhanced to support the evaluation of these devices.

A novel general method for reliability evaluation of fault-tolerant architectures, specifically enabling the introduction of fault tolerance, and the evaluation of circuit and architecture reliability is presented. The method is based on the modeling of probability density functions (PDFs) of unreliable components and their subsequent evaluation for a given reliability architecture. PDF modeling, presented for the first time in the context of realistic technology and arbitrary circuit size, is based on a cutting-edge reliability evaluation algorithm and offers scalability, speed, and accuracy. An estimation of the probability of failure of small to mid-sized circuits has been given with respect to the logic depth of these circuits. The evaluation of the effectiveness of analog averaging techniques in terms of reliability improvement compared to majority voter-based techniques has been performed, and it has been proven that the effectiveness significantly reduces with increased logic depth. Moreover, it has been shown that the local optimization can be efficiently performed by resynthesizing circuits, having the logic depth as the minimization goal.

A fast algorithm for reliability-optimal partitioning is proposed and has been realized. The algorithm thrives to solve the problem of combinational circuit partitioning for probability of failure minimization subject to area constraints. An analysis of fault-tolerant architectures in the context of the large-scale (trillion devices) system has been performed. Reliability-optimal redundancy factor and partition size as global parameters have been extracted in an analytical way. The optimal window of application of each fault-tolerant technique with respect to defect density has been derived. The analysis enables an optimum design trade-off between the reliability and power/area. R-fold modular redundancy with distributed voting and averaging voter has been selected as the most promising candidate for the implementation in trillion-transistor logic systems. Moreover, it has been demonstrated that a target reliability can be achieved with low to moderate redundancy factors ($R < 50$), even for high defect densities (device failure rate up to 10^{-3}).

A new fault-tolerant design methodology that integrates local-level evaluation, optimal reliability partitioning, and system-level reliability evaluation is presented. The methodology is envisioned as an "upgrade" of the existing digital design methodology and new concepts are merged into the methodology in a seamless and transparent way.

The proposed methodology has been demonstrated to offer a viable solution to the stringent issues raised by the massive defect densities that are expected to contaminate the early generations of nanoelectronic, as well as very deep-submicron CMOS-integrated circuits. Optimizations and adaptations are of course needed to improve the final fabrication yield and in-field operative time and map the specific constraints imposed by new fabrication technologies toward system design.

9.2 Conclusions or Back into the Big Picture

Building viable systems out of intrinsically unreliable atomic devices will have a high cost in terms of silicon area, and power dissipation, regardless of fault-tolerant techniques that are used in the implementations of complex microelectronic systems. Consequently, future nanoscale technologies must offer sufficient relaxation of design constraints to compensate for their unreliability. Improved fabrication technologies and fault-tolerant circuit techniques must be combined with higher level techniques, such as architectural, system, and even software techniques, in order to achieve an optimal level of reliability of the end-user electronic application. Totally fault-free operation may not be an achievable strategy.

Some fault-tolerant techniques may not be suitable to operate at any level of abstraction. Moreover, the required redundancy factors can be too large to support the development of meaningful applications. Dividing a system into different abstraction levels could improve the overall reliability, if the partitioning itself is performed in an optimal manner. Having this in mind, Fig. 8.22 illustrates an important conclusion. A priori system-level reliability evaluation and optimization is necessary to determine which is the optimal design space as well as its boundaries, where a given fault-tolerant technique can be applied. Moreover, the optimal partition size analysis, the reliability-optimal partitioning, and redundancy optimization are key steps in the successful realization of trillion devices system built out of highly unreliable components.

New methodologies and their supporting EDA tools are needed to respond to new requirements for realistic reliability evaluation and modeling by employing a data collection that does not oversimplify the models. This implies that the simulation methods also need to be flexible and support adjustable accuracy.

Finally, introducing reliability as a design parameter in future EDA tools as described in the presented design methodology enhancement should provide the designers the means of exploring the design space in terms of its reliability, in the similar sense as it is done today with respect to speed, power, and area constraints. Thus, finding the appropriate level of reliability in terms of a trade-off between technological constraints and application criticality may emerge as a realistic solution, in which microelectronic designers need to be supported by novel methods and tools, and which this book aspires to contribute to.

Appendix A
Probability of Chip and Signal Failure in System-Level Optimizations

This appendix presents specific derivations of the probability of chip failure for CRMR fault-tolerant architecture, as well as the derivation of the probability of signal failure in DRMR fault-tolerant architecture.

A.1 Probability of Chip Failure for Cascaded *R*-Fold Modular Redundancy Architecture

In derivation of probability of chip failure for the CRMR architecture the "zeroth layer" of CRMR (depicted in Fig. 8.15) can be equated with RMR architecture. Then expression (8.18) for the probability of chip failure in the case of RMR architecture and the optimal partition size can be written as

$$P_{\text{fails},0}^{\text{chip}} \approx \frac{N}{N_{\text{c}0}} F_{\text{out}} \left(P_{\text{fails},0}^{\text{red.unit}} + P_{\text{fails},0}^{\text{dec.gate}} \right) = \frac{N}{N_{\text{c}0}} F_{\text{out}} P_{\text{F}0}, \tag{A.1}$$

where $P_{\text{fails},0}^{\text{chip}}$ is the probability of failure of a chip, $P_{\text{fails},0}^{\text{red.unit}}$ is probability that R_0 redundant units fail, $P_{\text{fails},0}^{\text{dec.gate}}$ is the probability that decision gate fails, all of them given for the RMR ("zeroth-order" CRMR) fault-tolerant technique. $P_{\text{F}0}$ represents probability that an output of the reliable block ("zeroth layer") fails and, according to equivalent expression for RMR (8.18), is given as (considering $N_{\text{c,opt}}$)

$$P_{\text{F}0} = \left(\frac{1}{\frac{E_{1/2/3/}(R_0 + 1)}{2} - 1} + 1 \right) P_{\text{fails},0}^{\text{dec.gate}}. \tag{A.2}$$

Repeating this procedure for the "first-order" CRMR, the probability of chip failure is derived as

DOI 10.1007/978-1-4419-6217-1,

$$P_{\text{fails},1}^{\text{chip}} = \frac{N}{N_{c0}N_{c1}} F_{\text{out}} \left(P_{\text{fails},1}^{\text{red.u.}} + P_{\text{fails},1}^{\text{voter}} \right)$$
$$= \frac{N}{N_{c0}N_{c1}} F_{\text{out}} \left(\frac{1}{\dfrac{E_{1/2/3/}(R_1+1)}{2} - 1} + 1 \right) P_{\text{fails},1}^{\text{dec.gate}}, \tag{A.3}$$

where $P_{\text{fails},1}^{\text{chip}}$ is the probability of failure of a chip, $P_{\text{fails},1}^{\text{red.unit}}$ is the probability that R_1 redundant units fail, $P_{\text{fails},1}^{\text{dec.gate}}$ is the probability that a decision gate fails, and N_{c1} is the partition size; all given for "first layer" of CRMR. An assumption is that a unit output in "first layer" of CRMR (consisting of N_{c1} reliable blocks of "zeroth layer") fails if a corresponding output of any reliable blocks fail. Following this, it is also valid

$$P_{\text{fails},1}^{\text{unit}} = N_{c1} P_{F0}, \tag{A.4}$$

where P_{F0} represents the probability that an output of the reliable block in the "zeroth layer" fails. $N_{c1,\text{opt}}$ is derived in the same way as $N_{c,\text{opt}}$ in (8.17)

$$N_{c1,\text{opt}} = \frac{1}{4 \left. \dfrac{dP_{\text{fails},1}^{\text{unit}}}{dN_{c1}} \right|_{N_{c1}=N_{c1,\text{opt}}}} M(R_1)^{\dfrac{2}{E_{1/2/3}(R+1)}} = \frac{1}{4P_{F0}} M(R_1)^{\dfrac{2}{E_{1/2/3}(R+1)}},$$
$$M(R_i) = \left(\frac{1}{A\left(\dfrac{E_{1/2/3}(R_1+1)}{2} - 1 \right)} P_{\text{fails},i}^{\text{dec.gate}} \right). \tag{A.5}$$

The optimal partition size for a "zeroth layer" in CRMR is exactly the same size as the optimal partition size for RMR, as derived in (8.17), i.e., $N_{c0,\text{opt}} = N_{c,\text{opt}}$. This can be equivalently written as $N_{c0,\text{opt}} = mN_{c,\text{opt}}$, where $m = 1$, and therefore it will be proved that $m = 1$. Introducing the values of $N_{c0,\text{opt}}$ and $N_{c1,\text{opt}}$ as given in (8.17) and (A.5) into (A.3) yields an expression for $P_{\text{fails},1}^{\text{chip}}$ as follows:

$$P_{\text{fails},1}^{\text{chip}} = \frac{P_{F0}}{m} \frac{4^2 N \left. \dfrac{dP_{\text{fails}}^{\text{unit}}}{dN_c} \right|_{N_c=mN_{c,\text{opt}}}}{M(R_1)M(R_0)} F_{\text{out}} \left(\frac{1}{\dfrac{E_{1/2/3/}(R_1+1)}{2} - 1} + 1 \right) P_{\text{fails},1}^{\text{dec.gate}}. \tag{A.6}$$

After replacing N_c with $mN_{c,opt}$, following (8.13) $P_{fails,0}^{unit}$ becomes

$$P_{fails,0}^{unit}\Big|_{N_c=mN_{c,opt}} = \left.\frac{dP_{fails}^{unit}}{dN_c}\right|_{N_c=mN_{c,opt}} \cdot mN_{c,opt} \approx m \cdot P_{fails,0}^{unit}\Big|_{N_c=N_{c,opt}}. \quad (A.7)$$

Since probability of reliable block failure (at "zeroth layer") is given as

$$P_{F0} \approx A(4P_{fails}^{unit})^{E_{1/2/3}(R+1)/2} + P_{fails,0}^{dec.gate}, \quad (A.8)$$

after inserting (A.7) in (A.8), P_{F0} becomes

$$\begin{aligned} P_{F0} &\approx A\left(m(P_{fails,0}^{unit}\big|_{N_c=N_{c,opt}})\right)^{E_{1/2/3}(R+1)/2} + P_{fails,0}^{dec.gate} \\ &\approx \left(m^{E_{1/2/3}(R+1)/2}\frac{1}{\frac{E_{1/2/3}(R_0+1)}{2}-1}+1\right)P_{fails,0}^{dec.gate}. \end{aligned} \quad (A.9)$$

Finally, after inserting (A.7) in (A.6) $P_{fails,1}^{chip}$ becomes

$$\begin{aligned} P_{fails,1}^{chip} \approx \frac{1}{m}&\left(m^{E_{1/2/3}(R+1)/2}\frac{1}{\frac{E_{1/2/3}(R_0+1)}{2}-1}+1\right)P_{fails,0}^{dec.gate} \\ &\cdot\frac{4^2N\left.\frac{dP_{fails}^{unit}}{dN_c}\right|_{N_c=N_{c,opt}}}{M(R_1)M(R_0)}F_{out}\left(\frac{1}{\frac{E_{1/2/3}(R_1+1)}{2}-1}+1\right)P_{fails,1}^{dec.gate}. \end{aligned} \quad (A.10)$$

The expression $dP_{fails,1}^{chip}/dm = 0$ is equivalent to

$$\frac{d\left[\frac{1}{m}\left(m^{E_{1/2/3}(R+1)/2}\frac{1}{\frac{E_{1/2/3}(R_0+1)}{2}-1}+1\right)\right]}{dm} = 0 \quad (A.11)$$

that yields the optimum for $m = 1$, which proves that $N_{c0,opt} = N_{c,opt}$. For $m = 1$, (A.10) becomes

$$P_{\text{fails},1}^{\text{chip}} = \frac{4^2 N \left.\frac{dP_{\text{fails}}^{\text{unit}}}{dN_{\text{c}}}\right|_{N_{\text{c}}=N_{\text{c0,opt}}}}{M(R_1)M(R_0)} \left(\frac{1}{\frac{E_{1/2/3/}(R_1+1)}{2} - 1} + 1 \right) P_{\text{fails},1}^{\text{voter}} \cdot \left(\frac{1}{\frac{E_{1/2/3/}(R_0+1)}{2} - 1} + 1 \right) P_{\text{fails},0}^{\text{dec.gate}} \tag{A.12}$$

A.1.1 Generalization

Considering the general case of the "nth-order" CRMR, expression (A.12) transforms into

$$P_{\text{fails},n}^{\text{chip}} = N \left.\frac{dP_{\text{fails}}^{\text{unit}}}{dN_{\text{c}}}\right|_{N_{\text{c}}=N_{\text{c0,opt}}} \prod_{i=0}^{n} \frac{4}{M(R_i)} \left(\frac{1}{\frac{E_{1/2/3/}(R_i+1)}{2} - 1} + 1 \right) P_{\text{fails},i}^{\text{dec.gate}}. \tag{A.13}$$

Equation (A.2) is also generalized into

$$\begin{aligned} P_{\text{F}i} &= \left(\frac{1}{\frac{E_{1/2/3/}(R_i+1)}{2} - 1} + 1 \right) P_{\text{fails},i}^{\text{dec.gate}} \\ &= \left(\frac{1}{\frac{E_{1/2/3/}(R_i+1)}{2} - 1} + 1 \right) (m+2R)p_{\text{f}} \end{aligned} \tag{A.14}$$

and

$$N_{\text{c}i+1} = \frac{M(R_{i+1})}{2P_{\text{F}i}}, \tag{A.15}$$

where $i \in [0,\ n-1]$. Finally, by inserting (A.14) into (A.13)

$$P_{\text{fails},n}^{\text{chip}} = f(R_0, R_1, \ldots, R_n) = N \left.\frac{dP_{\text{fails}}^{\text{unit}}}{dN_{\text{c}}}\right|_{N_{\text{c}}=N_{\text{c0,opt}}} \prod_{i=0}^{n} \frac{4P_{\text{F}i}}{M(R_i)}. \tag{A.16}$$

A.2 Probability of Input Signals Failure in Distributed *R*-Fold Modular Redundancy Architecture

In order to calculate the probability that input signals to i out of R redundant units in DRMR fault-tolerant architecture are erroneous ($P^{\text{in.sig.}}_{\text{fails},(i)}$) let us consider different cases depending on the number of units (i) failing due to input signals:

Case 1:

All input signals of all redundant units in the reliable block are fault free. This is equivalent to all fault-free decision gates driving all the signals for all redundant units. The probability of this event to occur is

$$P^{\text{in.sig.}}_{\text{fails},(0)} = (1 - P^{\text{dec.gate}}_{\text{fails}})^{RF_{\text{in}}}, \tag{A.17}$$

where $P^{\text{dec.gate}}_{\text{fails}}$ is the probability of decision gate failure given in (8.8) in the list of main assumptions in Section 8.3 and F_{in} is the number of inputs to each redundant unit (fanin).

The probability of an output of the reliable block to fail when zero units fail due to input signals ($P^{\text{block}}_{\text{fails},(0)}$) is given as

$$\begin{aligned} P^{\text{block}}_{\text{fails},(0)} &= P^{\text{red.unit}}_{\text{fails},(1)} \cdot P^{\text{in.sig.}}_{\text{fails},(1)} \\ &= 2\binom{R}{\frac{R+1}{2}}(1 - P^{\text{unit}}_{\text{fails}})^{(R-3)/2}(P^{\text{unit}}_{\text{fails}})^{E_{1/2/3}\frac{R+1}{2}}(1 - P^{\text{dec.gate}}_{\text{fails}})^{RF_{\text{in}}}. \end{aligned} \tag{A.18}$$

Case 2:

One or more input signals to exactly one redundant unit in the reliable block are faulty. This is equivalent to failure of some decision gates driving signals to exactly one redundant unit. Decision gates that are driving all signals to all other redundant units have to be fault free. The probability of this event to occur is

$$P^{\text{in.sig.}}_{\text{fails},(1)} = R\left[1 - (1 - P^{\text{dec.gate}}_{\text{fails}})^{F_{\text{in}}}\right](1 - P^{\text{dec.gate}}_{\text{fails}})^{(R-1)F_{\text{in}}}, \tag{A.19}$$

where $P^{\text{dec.gate}}_{\text{fails}}$ is the probability of decision gate failure and F_{in} is the number of inputs to each redundant unit (fanin).

The probability of an output of the reliable block to fail when one single unit fails due to input signals ($P^{\text{block}}_{\text{fails},(1)}$) is given as

$$
\begin{aligned}
P^{\text{block}}_{\text{fails},(1)} &= P^{\text{red.unit}}_{\text{fails},(1)} P^{\text{in.sig.}}_{\text{fails},(1)} \\
&= 2\binom{R-1}{\frac{R+1}{2}-1}(1-P^{\text{unit}}_{\text{fails}})^{(R-3)/2}(P^{\text{unit}}_{\text{fails}})^{E_{1/2/3}\frac{R-1}{2}} \\
&\quad \cdot R\left[1-(1-P^{\text{dec.gate}}_{\text{fails}})^{RF_{\text{in}}}\right](1-P^{\text{dec.gate}}_{\text{fails}})^{(R-1)F_{\text{in}}}
\end{aligned}
\tag{A.20}
$$

Case n (general case):

One or more input signals to exactly n redundant units (out of R in total) in the reliable block are faulty. This is equivalent to failure of some decision gates driving signals to exactly n redundant units. Decision gates that are driving all signals to all other redundant units have to be fault free. The probability of this event to occur is expressed as

$$
P^{\text{in.sig.}}_{\text{fails},(n)} = \binom{R}{n}\left[1-(1-P^{\text{dec.gate}}_{\text{fails}})^{F_{\text{in}}}\right]^n (1-P^{\text{dec.gate}}_{\text{fails}})^{(R-n)F_{\text{in}}}, \tag{A.21}
$$

where $P^{\text{dec.gate}}_{\text{fails}}$ is the probability of decision gate failure and F_{in} is the number of inputs to each redundant unit (fanin). Here it is assumed that $P^{\text{dec.gate}}_{\text{fails}} < 1\%$ and the approximation $(1-x)^n \approx 1-nx$ is justified. With this approximation (A.21) becomes

$$
P^{\text{in.sig.}}_{\text{fails},(n)} = \binom{R}{n}(F_{\text{in}}P^{\text{dec.gate}}_{\text{fails}})^n(1-P^{\text{dec.gate}}_{\text{fails}})^{(R-n)F_{\text{in}}}, \tag{A.22}
$$

The probability of an output of the reliable block to fail when n units fail due to input signals ($P^{\text{block}}_{\text{fails},(n)}$) is given as

$$
\begin{aligned}
P^{\text{block}}_{\text{fails},(n)} &= P^{\text{red.unit}}_{\text{fails},(n)} \cdot P^{\text{in.sig.}}_{\text{fails},(n)} \\
&= 2\binom{R-n}{\frac{R+1}{2}-n}(1-P^{\text{unit}}_{\text{fails}})^{(R-3)/2}(P^{\text{unit}}_{\text{fails}})^{E_{1/2/3}\left(\frac{R+1}{2}-n\right)} \\
&\quad \cdot \binom{R}{n}(F_{\text{in}}P^{\text{dec.gate}}_{\text{fails}})^n(1-P^{\text{dec.gate}}_{\text{fails}})^{(R-n)F_{\text{in}}} \\
&= 2\binom{R}{\frac{R+1}{2}}\binom{\frac{R+1}{2}}{n}(1-P^{\text{unit}}_{\text{fails}})^{(R-3)/2}(1-P^{\text{dec.gate}}_{\text{fails}})^{\frac{R-1}{2}F_{\text{in}}} \\
&\quad \cdot (P^{\text{unit}}_{\text{fails}})^{E_{1/2/3}\left(\frac{R+1}{2}-n\right)}(1-P^{\text{dec.gate}}_{\text{fails}})^{\left(\frac{R+1}{2}-n\right)F_{\text{in}}}(F_{\text{in}}P^{\text{dec.gate}}_{\text{fails}})^n
\end{aligned}
\tag{A.23}
$$

The equality $\binom{R-n}{\frac{R+1}{2}-n}\binom{R}{n} = \binom{R}{\frac{R+1}{2}}\binom{\frac{R+1}{2}}{n}$ is used in (A.23).

References

1. G. E. Moore, "Cramming more components onto integrated circuits," *Electronics Magazine*, vol. 38, pp. 114–117, 1965.
2. (2007) International technology roadmap for semiconductors. [Online]. Available: http://www.itrs.net/Links/2007ITRS/Home2007.htm
3. L. W. Counts, "Analog and mixed-signal innovation: The process-circuit-system-application interaction," in *Proceedings of Digest of Technical Papers. IEEE International Solid-State Circuits Conference ISSCC 2007*, 2007, pp. 24–30.
4. C. Constantinescu, "Trends and challenges in VLSI circuit reliability," *IEEE Micro*, vol. 23, no. 4, pp. 14–19, July–Aug. 2003.
5. J. R. Heath, P. J. Kuekes, G. S. Snider, and R. S. Williams, "A defect-tolerant computer architecture: Opportunities for nanotechnology," *Science*, vol. 280, pp. 1716–1721, 1998.
6. K. Nikolic, A. Sadek, and M. Forshaw, "Fault-tolerant techniques for nanocomputers," *Nanotechnology*, vol. 13, pp. 357–362, 2002.
7. K. K. Likharev, "Single-electron devices and their applications," *Proceedings of the IEEE*, vol. 87, no. 4, pp. 606–632, April 1999.
8. U. Feldkamp and C. M. Niemeyer, "Rational design of DNA nanoarchitectures," *Angewandte Chemie International Edition*, vol. 45, pp. 1856–1876, 2006.
9. C. Lin, Y. Liu, S. Rinker, and H. Yan, "DNA tile based self-assembly: Building complex nanoarchitectures," *Chemical Physics and Physical Chemistry*, vol. 7, no. 8, pp. 1641–1647, 2006.
10. J. E. Green, J. W. Choi, A. Boukai, Y. Bunimovich, E. Johnston-Halperin, E. Delonno, Y. Luo, B. A. Sheriff, K. Xu, Y. S. Shin, H.-R. Tseng, J. F. Stoddart, and J. R. Heath, "A 160-kilobit molecular electronic memory patterned at 10^{11} bits per square centimeter," *Nature*, vol. 445, pp. 414–417, 2007.
11. *Global System IC (ASSP/ASIC) Service Management Report*, International Business Solutions, May 2004.
12. (2009) International technology roadmap for semiconductors. [Online]. Available: http://www.itrs.net/Links/2009ITRS/Home2009.htm
13. J. von Neumann, *Automata Studies*. Princeton, NJ: Princeton University Press, 1956, ch. Probabilistic logic and the synthesis of reliable organisms from unreliable components, pp. 43–98.
14. S. Cotofana, A. Schmid, Y. Leblebici, A. M. Ionescu, O. Soffke, P. Zipf, M. Glesner, and A. Rubio, "CONAN - a design exploration framework for reliable nano-electronics," in *Proceedings of the 16th IEEE International Conference on Application-Specific Systems, Architecture Processors (ASAP)*, 23–25 July 2005, pp. 260–267.
15. S. K. Shukla, R. Karri, S. C. Goldstein, F. Brewer, K. Banerjee, and S. Basu, "Nano, quantum, and molecular computing: Are we ready for the validation and test challenges?" in *Proceedings of the 8th IEEE International High-Level Design Validation and Test Workshop*, 2003, pp. 3–7.

16. S. Lazarova-Molnar, V. Beiu, and W. Ibrahim, "Reliability the fourth optimization pillar of nanoelectronics," in *Proceedings IEEE International Conference on Signal Processing and Communications (ICSPC)*, 24–27 Nov. 2007, pp. 73–76.
17. M. L. Bushnel and W. D. Agrawal, *Essentials of Electronic Testing for Digital, Memory, and Mixed-Signal VLSI Circuits*. Boston, MA: Springer, 2005.
18. B. Dhillon, *Design Reliability: Fundamentals and Applications*. Boca Raton, FL: CRC Press, 1999.
19. B. Dhillon, *Reliability, Quality, and Safety for Engineers*. Boca Raton, FL: CRC Press, 2005.
20. E. Hnatek, *Practical Reliability of Electronic Equipment and Products*. Boca Raton, FL: CRC Press, 2003.
21. L.-T. Wang and C. E. Stroud, *Power-Aware Testing and Test Strategies for Low Power Devices*. Boston, MA: Springer, 2010, ch. Fundamentals of VLSI testing, pp. 1–29. [Online]. Available: http://www.springerlink.com/content/978-1-4419-0927-5
22. M. J. Howes and D. Morge, *Reliability and Degradation - Semiconductor Devices and Circuits*, M. J. Howes and D. Morge, (eds.). Chichester: Wiley-Interscience, 1981.
23. C. Chiang and K. Jamil, *Design for Manufacturability and Yield for Nano-Scale Cmos*, ser. Integrated Circuits and Systems. Dordrecht: Springer, 2007. [Online]. Available: http://www.springerlink.com/content/978-1-4020-5187-6
24. R. Rodríguez-Montañés, E. M. J. G. Bruls, and J. Figueras, "Bridging defects resistance in the metal layer of a CMOS process," *Journal of Electronic Testing*, vol. 8, no. 1, pp. 35–46, Feb. 1996. [Online]. Available: http://www.springerlink.com/content/k5833m20t1131v46/
25. J. Arlat and Y. Crouzet, "Physical fault models and fault tolerance," in *Models in Hardware Testing*, ser. Frontiers in Electronic Testing, vol. 43. Dordrecht: Springer, 2010, pp. 217–255. [Online]. Available: http://www.springerlink.com/content/w047645173x25605/
26. T. Lehtonen, J. Plosila, and J. Isoaho, "On fault tolerance techniques towards nanoscale circuits and systems," Turku Center for CS, University of Turku, Turku, Finland, Technical Report 708, 2005.
27. M. Pflanz, *On-line Error Detection and Fast Recover Techniques for Dependable Embedded Processors*. Berlin: Springer, 2002, vol. 2270, ch. Fault models and fault-behavior of processor structures.
28. W. Heidergott, "SEU tolerant device, circuit and processor design," in *Proceedings of the 42nd Design Automation Conference (DAC)*, 13–17 June 2005, pp. 5–10.
29. R. D. Eldred, "Test routines based on symbolic logical statements," *Journal of ACM*, vol. 6, no. 1, pp. 33–37, 1959.
30. J. A. B. Fortes, "Future challenges in VLSI system design," in *Proceedings of IEEE Computer Society Annual Symposium on VLSI*, 20–21 Feb. 2003, pp. 5–7.
31. V. Beiu, W. Ibrahim, and S. Lazarova-Molnar, "A fresh look at majority multiplexing when devices get into the picture," in *Proceedings of the 7th IEEE Conference on Nanotechnology (IEEE-NANO)*, 2–5 Aug. 2007, pp. 883–888.
32. J. P. Shen, W. Maly, and F. J. Ferguson, "Inductive fault analysis of MOS integrated circuits," *IEEE Design & Test of Computers*, vol. 2, no. 6, pp. 13–26, Dec. 1985.
33. T. Olbrich, J. Perez, I. A. Grout, A. M. D. Richardson, and C. Ferrer, "Defect-oriented vs schematic-level based fault simulation for mixed-signal ICs," in *Proceedings of the International Test Conference (ITC)*, 20–25 Oct. 1996, pp. 511–520.
34. D. Al-Khalili, S. Adham, C. Rozon, M. Hossain, and D. Racz, "Comprehensive defect analysis and defect coverage of CMOS circuits," in *Proceedings of the IEEE International Symposium on Defect and Fault Tolerance in VLSI Systems (DFT)*, 2–4 Nov. 1998, pp. 84–92.
35. R. Rodriguez-Montanes, J. A. Segura, V. H. Champac, J. Figueras, and J. A. Rubio, "Current vs. logic testing of gate oxide short, floating gate short and bridging failures in CMOS," in *Proceedings of the International Test Conference (ITC)*, 26–30 Oct 1991, p. 510.
36. T. M. Storey and W.Maly, "CMOS bridging faults," in *Proceedings of the International Test Conference (ITC)*, 1991, pp. 1123–1131.

37. R. Rodriguez-Montanes, E. M. J. G. Bruis, and J. Figueras, "Bridging defects resistance measurements in a CMOS process," in *Proceedings of the International Test Conference (ITC)*, Sept. 20–24 1992, p. 892.
38. M. Sytrzycki, "Modeling of gate oxide shorts in MOS transistors," *IEEE Transactions on Computer-Aided Design of Integrated Circuits and Systems*, vol. 8, no. 3, pp. 193–202, Mar. 1989.
39. A. Cabrini, D. Cantarelli, P. Cappelletti, R. Casiraghi, A. Maurelli, M. Pasotti, P. L. Rolandi, and G. Torelli, "A test structure for contact and via failure analysis in deep-submicrometer CMOS technologies," *IEEE Transactions on Semiconductor Manufacturing*, vol. 19, no. 1, pp. 57–66, Feb. 2006.
40. S. Mukhopadhyay, K. Kim, K. A. Jenkins, C.-T. Chuang, and K. Roy, "An on-chip test structure and digital measurement method for statistical characterization of local random variability in a process," *IEEE Journal of Solid-State Circuits*, vol. 43, no. 9, pp. 1951–1963, Sept. 2008.
41. V. Krishnaswamy, A. B. Ma, and P. Vishakantaiah, "A study of bridging defect probabilities on a Pentium (TM) 4 CPU," in *Proceedings of the International Test Conference (ITC)*, 30 Oct.–1 Nov. 2001, pp. 688–695.
42. V. Beiu, "Grand challenges of nanoelectronics and possible architectural solutions: What do Shannon, von Neumann, Kolmogorov, and Feynman have to do with Moore," in *Proceedings of the 37th International Symposium on Multiple-Valued Logic (ISMVL)*, 13–16 May 2007, p. 1–6.
43. J. A. Hutchby, G. I. Bourianoff, V. V. Zhirnov, and J. E. Brewer, "Extending the road beyond CMOS," *IEEE Circuits and Devices Magazine*, vol. 18, no. 2, pp. 28–41, 2002.
44. R. H. Chen, A. N. Korotkov, and K. K. Likharev, "Single-electron transistor logic," *Applied Physics Letters*, vol. 68, pp. 1954–1956, 1996.
45. C. P. Heij, P. Hadley, and J. E. Mooij, "Single-electron inverter," *Applied Physics Letters*, vol. 78, pp. 1140–1142, 2001.
46. Y. Ono, Y. Takahashi, K. Yamazaki, M. Nagase, H. Namatsu, K. Kurihara, and K. Murase, "Si complementary single-electron inverter with voltage gain," *Applied Physics Letters*, vol. 76, pp. 3121–3123, 2000.
47. S. Mahapatra, A. M. Ionescu, and K. Banerjee, "A quasi-analytical SET model for few electron circuit simulation," *IEEE Electron Device Letters*, vol. 23, no. 6, pp. 366–368, June 2002.
48. K. Uchida, J. Koga, R. Ohba, and A. Toriumi, "Programmable single-electron transistor logic for future low-power intelligent LSI: Proposal and room-temperature operation," *IEEE Transactions on Electron Devices*, vol. 50, no. 7, pp. 1623–1630, July 2003.
49. N. Asahi, M. Akazawa, and Y. Amemiya, "Single-electron logic systems based on the binary decision diagram," *IEICE Transactions on Electronics*, vol. E81-C, pp. 49–56, 1998.
50. S. Kasai and H. Hasegawa, "GaAs and InGaAs single electron hexagonal nanowire circuits based on binary decision diagram logic architecture," *Physica E*, vol. 13, pp. 925–929, 2002.
51. Y. Takahashi, A. Fujiwara, Y. Ono, and K. Murase, "Silicon single-electron devices and their applications," in *Proceedings of the 30th IEEE International Symposium on Multiple-Valued Logic (ISMVL)*, 23–25 May 2000, pp. 411–420.
52. Z. A. K. Durrani, A. C. Lnine, and H. Ahmed, "Coulomb blockade memory using integrated single-electron transistor/metal-oxide-semiconductor transistor gain cells," *IEEE Transactions on Electron Devices*, vol. 47, no. 12, pp. 2334–2339, Dec. 2000.
53. H. Mizuta, H. O. Muller, K. Tsukagoshi, D. Williams, Z. Durrani, A. Irvine, G. Evans, S. Amakawa, K. Nakazato, and H. Ahmed, "Nanoscale Coulomb blockade memory and logic devices," *Nanotechnology*, vol. 12, no. 2, pp. 155–159, 2001.
54. N. J. Stone, H. Ahmed, and K. Nakazato, "A high-speed silicon single-electron random access memory," *IEEE Electron Device Letters*, vol. 20, no. 11, pp. 583–585, Nov. 1999.
55. J. Han and P. Jonker, "A system architecture solution for unreliable nanoelectronic devices," *IEEE Transactions on Nanotechnology*, vol. 1, no. 4, pp. 201–208, Dec. 2002.

56. (2000, Nov.) Technology roadmap for nanoelectronics. Microelectronics Advanced Research Initiative - MELARI NANO. [Online]. Available: http://www.itrs.net/links/2001itrs/Links/modeling/Nano2000WEBVersion.pdf
57. P. Mazumder, S. Kulkarni, M. Bhattacharya, J. P. Sun, and G. I. Haddad, "Digital circuit applications of resonant tunneling devices," *Proceedings of the IEEE*, vol. 86, no. 4, pp. 664–686, April 1998.
58. J. P. A. van der Wagt, A. C. Seabaugh, and I. Beam, E. A., "RTD/HFET low standby power SRAM gain cell," *IEEE Electron Device Letters*, vol. 19, no. 1, pp. 7–9, Jan. 1998.
59. M. A. Reed, W. R. Frensley, R. J. Matyi, J. N. Randall, and A. C. Seabaugh, "Realization of a three-terminal resonant tunneling device: The bipolar quantum resonant tunneling transistor," *Applied Physics Letters*, vol. 54, pp. 1034–1036,, 1989.
60. J. Stock, J. Malindretos, K. M. Indlekofer, M. Pottgens, A. Forster, and H. Luth, "A vertical resonant tunneling transistor for application in digital logic circuits," *IEEE Transactions on Electron Devices*, vol. 48, no. 6, pp. 1028–1032, June 2001.
61. V. V. Zhirnov, J. A. Hutchby, G. I. Bourianoffls, and J. E. Brewer, "Emerging research logic devices," *IEEE Circuits and Devices Magazine*, vol. 21, no. 3, pp. 37–46, 2005.
62. E. F. Codd, *Cellular Automata*. London: Academic Press, 1968.
63. C. S. Lent and P. D. Tougaw, "A device architecture for computing with quantum dots," *Proceedings of the IEEE*, vol. 85, no. 4, pp. 541–557, April 1997.
64. G. Bourianoff, "The future of nanocomputing," *Computer*, vol. 36, no. 8, pp. 44–53, Aug. 2003.
65. A. O. Orlov, I. Amlani, R. Kummamuru, R. Ramasubramaniam, G. Toth, C. S. Lent, G. H. Bernstein, and G. L. Snider, "Experimental demonstration of clocked single-electron switching in quantum-dot cellular automata," *Applied Physics Letters*, vol. 77, pp. 295–297, 2000.
66. A. O. Orlov, R. Kummamuru, R. Ramasubramaniam, G. Toth, C. S. Lent, G. H. Bernstein, and G. L. Snider, "Experimental demonstration of a latch in clocked quantum-dot cellular automata," *Applied Physics Letters*, vol. 78, pp. 1625–1627, 2001.
67. A. O. Orlov, R. Kummamuru, R. Ramasubramaniam, G. Toth, C. S. Lent, G. H. Bernstein, and G. L. Snider, "Clocked quantum-dot cellular automata shift register," *Surface Science*, vol. 532, pp. 1193–1198, 2003.
68. P. D. Tougaw and C. S. Lent, "Dynamic behavior of quantum cellular automata," *Journal of Applied Physics*, vol. 80, pp. 4722–4736, 1996.
69. C. S. Lent, B. Isaksen, and M. Lieberman, "Molecular quantum-dot cellular automata," *Journal of American Chemical Society*, vol. 125, pp. 1056–1063, 2003.
70. K. Nikolic, D. Berzon, and M. Forshaw, "Relative performance of three nanoscale devices - CMOS, RTDs and QCAs - against a standard computing task," *Nanotechnology*, vol. 12, no. 1, pp. 38–43, 2001.
71. R. P. Cowburn and M. E. Welland, "Room temperature magnetic quantum cellular automata," *Science*, vol. 287, no. 5457, pp. 1466–1468, 2000.
72. D. A. Allwood, G. Xiong, M. D. Cooke, C. C. Faulkner, D. Atkinson, N. Vernier, and R. P. Cowburn, "Submicrometer ferromagnetic NOT gate and shift register," *Science*, vol. 296, pp. 2003–2006, 2002.
73. K. B. K. Teo, R. G. Lacerda, M. H. Yang, A. S. Teh, L. A. W. Robinson, S. H. Dalal, N. L. Rupesinghe, M. Chhowalla, S. B. Lee, D. A. Jefferson, D. G. Hasko, G. A. J. Amaratunga, W. L. Milne, P. Legagneux, L. Gangloff, E. Minoux, J. P. Schnell, and D. Pribat, "Carbon nanotube technology for solid state and vacuum electronics," *IEE Proceedings -Circuits, Devices and Systems*, vol. 151, no. 5, pp. 443–451, 2004.
74. S. J. Tans, A. R. M. Verschueren, and C. Dekker, "Room-temperature transistor based on a single carbon nanotube," *Nature*, vol. 393, pp. 49–52, 1998.
75. H. W. C. Postma, T. F. Teepen, Z. Yao, M. Grifoni, and C. Dekker, "Carbon nanotubes single-electron transistors at room temperature," *Science*, vol. 293, pp. 76–79, 2001.
76. A. Bachtold, P. Hadley, T. Nakanishi, and C. Dekker, "Logic circuits with carbon nanotube transistors," *Science*, vol. 294, pp. 1317–1320, 2001.

77. A. Javey, Q. Wang, A. Urai, Y. Li, and H. Dai, "Carbon nanotube transistor arrays for multistage complementary logic and ring oscillators," *Nano Letters*, vol. 2, pp. 929–932, 2002.
78. Y. Huang, X. F. Duan, Q. Wei, and C. M. Lieber, "Directed assembly of one-dimensional nanostructures into functional networks," *Science*, vol. 291, pp. 630–633, 2001.
79. N. A. Melosh, A. Boukai, F. Diana, B. Gerardot, A. Badolato, P. M. Petroff, and J. R. Heath, "Ultrahigh-density nanowire lattices and circuits," *Science*, vol. 300, pp. 112–115, 2003.
80. X. F. Duan, Y. Huang, Y. Cui, J. F. Wang, and C. M. Lieber, "Indium phosphide nanowires as building blocks for nanoscale electronic and optoelectronic devices," *Nature*, vol. 409, no. 6816, pp. 66–69, 2001.
81. Y. Huang, X. Duan, Y. Cui, L. J. Lauhon, K. Kim, and C. M. Lieber, "Logic gates and computation from assembled nanowire building blocks," *Science*, vol. 294, pp. 1313–1317, 2001.
82. Z. Zhong, D. Wang, Y. Cui, M. W. Bockrath, and C. M. Lieber, "Nanowire crossbar arrays as address decoders for integrated nanosystems," *Science*, vol. 302, pp. 1377–1379, 2003.
83. C. Joachim, J. K. Gimzewski, and A. Aviram, "Electronics using hybrid-molecular and mono-molecular devices," *Nature*, vol. 408, pp. 541–548, 2000.
84. H. Park, J. Park, A. K. L. Lim, E. H. Anderson, A. P. Alivisatos, and P. L. McEuen, "Nanomechanical oscillations in a single-C-60 transistor," *Nature*, vol. 407, pp. 57–60, 2000.
85. Y. Luo, C. P. Collier, J. O. Jeppesen, K. A. Nielsen, E. Delonno, G. Ho, J. Perkins, H. R. Tseng, T. Yamamoto, J. F. Stoddart, and J. R. Heath, "Two-dimensional molecular electronics circuits," *Chemical Physics and Physical Chemistry* , vol. 3, no. 6, pp. 519–525, 2002.
86. M. R. Stan, P. D. Franzon, S. C. Goldstein, J. C. Lach, and M. M. Ziegler, "Molecular electronics: From devices and interconnect to circuits and architecture," *Proceedings of the IEEE*, vol. 91, no. 11, pp. 1940–1957, 2003.
87. X. Ma, D. B. Strukov, J. H. Lee, and K. K. Likharev, "Afterlife for silicon: Cmol circuit architectures," in *Proceedings of the 5th IEEE Conference Nanotechnology*, 2005, pp. 175–178.
88. P. Bunyk, K. Likharev, and D. Zinoviev, "RSFQ technology: Physics and devices," *International Journal of High Speed Electronics and Systems*, vol. 11, no. 1, pp. 257–305, 2001.
89. W. Chen, A. V. Rylyakov, V. Patel, J. E. Lukens, and K. K. Likharev, "Rapid single flux quantum T-flip flop operating up to 770 GHz," *IEEE Transactions on Applied Superconductivity*, vol. 9, no. 2, pp. 3212–3215, 1999.
90. D. K. Brock, "RSFQ technology: Circuits and systems," *International Journal of High Speed Electronics*, vol. 11, no. 1, pp. 307–362, 2001.
91. A. M. Kadin, C. A. Mancini, M. J. Feldman, and D. K. Brock, "Can RSFQ logic circuits be scaled to deep submicron junctions?" *IEEE Transactions on Applied Superconductivity*, vol. 11, no. 1, pp. 1050–1055, 2001.
92. D. K. Brock, E. K. Track, and J. M. Rowell, "Superconductor ICs: The 100-GHz second generation," *IEEE Spectrum*, vol. 37, no. 12, pp. 40–46, 2000.
93. J. E. Mooij, T. P. Orlando, L. Levitov, L. Tian, C. H. van der Wal, and S. Lloyd, "Josephson persistent-current qubit," *Science*, vol. 285, pp. 1036–1039, 1999.
94. I. Chiorescu, Y. Nakamura, C. J. P. M. Harmans, and J. E. Mooij, "Coherent quantum dynamics of a superconducting flux qubit," *Science*, vol. 299, pp. 1869–1871, 2003.
95. P. Jonker and J. Han, "On quantum and classical computing with arrays of superconducting persistent current qubits," in *Proceedings of the 5th IEEE International Workshop on Computer Architectures for Machine Perception*, 2000, pp. 69–78.
96. J. Han and P. Jonker, "Novel computing architecture on arrays of Josephson persistent current bits," in *Proceedings of the 5thth International Conference on Modeling and Simulation of Microsystems (MSM)*, San Juan, Puerto Rico, USA, April 2002, pp. 636–639.
97. M. Johnson, "The all-metal spin transistor," *IEEE Spectrum*, vol. 31, no. 5, pp. 47–51, 1994.
98. M. Johnson, "The bipolar spin transistor. A novel solid state device taking its first steps," *IEEE Potentials*, vol. 14, no. 1, pp. 26–30, 1995.

99. S. Sugahara, "Spin MOSFETs as a basis for integrated spin-electronics," in *Proceedings of the 5th IEEE Conference on Nanotechnology (IEEE-NANO)*, vol. 1, 2005, pp. 142–145.
100. C. L. Dennis, C. V. Tiusan, J. F. Gregg, G. J. Ensell, and S. M. Thompson, "Silicon spin diffusion transistor: Materials, physics and device characteristics," *IEE Proceedings -Circuits, Devices and Systems*, vol. 152, no. 4, pp. 340–354, 2005.
101. M. Johnson, "Magnetoelectronic memories last and last," *IEEE Spectrum*, vol. 37, no. 2, pp. 33–40, 2000.
102. J. Lille, N. Smith, and B. A. Gurney, "Requirements for integration of a magnetic transistor into a read head," in *Proceedings of the IEEE International Magnetics Conference (INTERMAG)*, 2006, pp. 489–489.
103. C. K. Lo, Y. W. Huang, Y. D. Yao, D. R. Huang, and J. H. Huang, "Spin transistor for magnetic recording," *IEEE Transactions on Magnetics*, vol. 41, no. 2, pp. 892–895, 2005.
104. M. Tanaka and S. Sugahara, "MOS-based spin devices for reconfigurable logic," *IEEE Transactions on Electron Devices*, vol. 54, no. 5, pp. 961–976, 2007.
105. M. Forshaw, D. Crawley, P. Jonker, J. Han, and C. S. Torres, "Nano Arch review: A review of the status of research and training into architectures for nanoelectronic and nanophotonic systems in the European research area," FP6/2002/IST/1 (Ext. rep. 507519). London: University College London., Technical Report, 2004.
106. V. Beiu, S. Aunet, J. Nyathi, R. R. Rydberg, and W. Ibrahim, "Serial addition: Locally connected architectures," *IEEE Transactions on Circuits and Systems I: Regular Papers*, vol. 54, no. 11, pp. 2564–2579, Nov. 2007.
107. M. Hartmann and P. C. Haddow, "Evolution of fault-tolerant and noise-robust digital designs," *IEE Proceedings -Computers and Digital Techniques*, vol. 151, no. 4, pp. 287–294, 2004.
108. M. Forshaw, R. Stadler, D. Crawley, and K. Nicolic, "A short review of nanoelectronic architectures," *Nanotechnology*, vol. 15, pp. S220–S223, 2004.
109. S. Lazarova-Molnar, V. Beiu, and W. Ibrahim, "A strategy for reliability assessment of future nano-circuits," in *Proceedings of the 11th WSEAS International Conference on Circuits (ICC)*. Stevens Point, WI: World Scientific and Engineering Academy and Society (WSEAS), 2007, pp. 60–65.
110. D. P. Siewiorek and R. S. Swarz, *The Theory and Practice of Reliable System Design*. Bedford, MA: Digital Press, 1982.
111. P. G. Depledge, "Fault-tolerant computer systems," *IEE Proceedings A Physical Science, Measurement and Instrumentation, Management and Education, Reviews*, vol. 128, no. 4, pp. 257–272, May 1981.
112. J. A. Abraham and D. P. Siewiorek, "An algorithm for the accurate reliability evaluation of triple modular redundancy networks," *IEEE Transactions on Computers*, vol. 1, no. 7, pp. 682–692, July 1974.
113. B. W. Johnson, *Design and Analysis of Fault-Tolerant Digital Systems*. Reading, MA: Addison-Wesley Publishing Company, 1989.
114. S. Spagocci and T. Fountain, "Fault rates in nanochip devices," *Proceedings of the Electrochemical Society*, vol. 98, no. 19, 1999, pp. 582–593.
115. M. Stanisavljević, A. Schmid, and Y. Leblebici, "Optimization of nanoelectronic systems' reliability under massive defect density using cascaded R-fold modular redundancy," *Nanotechnology*, vol. 19, no. 46, pp. 1–9, 2008.
116. R. I. Bahar, J. Chen, and J. Mundy, *Nano, Quantum and Molecular Computing: Implications to High Level Design and Validation*. Norwell, MA: Kluwer Academic Publishers, 2004, ch. Nanocomputing in the presence of defects and faults: A survey, pp. 39–72.
117. W. H. Pierce, *Failure-Tolerant Computer Design*. New York, NY: Academic Press, 1965.
118. J. Han and P. Jonker, "From massively parallel image processors to fault-tolerant nanocomputers," in *Proceedings of the 17th International Conference on Pattern Recognition (ICPR)*, vol. 3, 23–26 Aug. 2004, pp. 2–7.

119. J. Han, J. Gao, P. Jonker, Y. Qi, and J. A. B. Fortes, "Toward hardware-redundant, fault-tolerant logic for nanoelectronics," *IEEE Design & Test of Computers*, vol. 22, no. 4, pp. 328–339, July–Aug. 2005.
120. J. Tryon, *Redundancy Techniques for Computing Systems*. Washington, DC: Spartan Books, 1962, ch. Quadded logic, pp. 205–228.
121. S. Roy and V. Beiu, "Multiplexing schemes for cost-effective fault-tolerance," in *Proceedings of the 4th IEEE Conference on Nanotechnology (IEEE-NANO)*, 16–19 Aug. 2004, pp. 589–592.
122. S. Roy and V. Beiu, "Majority multiplexing-economical redundant fault-tolerant designs for nanoarchitectures," *IEEE Transactions on Nanotechnology*, vol. 4, no. 4, pp. 441–451, July 2005.
123. A. S. Sadek, K. Nikolic, and M. Forshaw, "Parallel information and computation with restitution for noise-tolerant nanoscale logic networks," *Nanotechnology*, vol. 15, pp. 192–210, 2004.
124. J. Han and P. Jonker, "A defect- and fault-tolerant architecture for nanocomputers," *Nanotechnology*, vol. 14, pp. 224–230, 2003.
125. G. Norman, D. Parker, M. Kwiatkowska, and S. Shukla, "Evaluating the reliability of NAND multiplexing with PRISM," *IEEE Transactions on Computer-Aided Design of Integrated Circuits and Systems*, vol. 24, no. 10, pp. 1629–1637, Oct. 2005.
126. G. Roelke, R. Baldwin, and D. Bulutoglu, "Analytical models for the performance of von Neumann multiplexing," *IEEE Transactions on Nanotechnology*, vol. 6, no. 1, pp. 75–89, Jan. 2007.
127. M. Forshaw, K. Nikolic, and A. S. Sadek, "ANSWERS: Autonomous nanoelectronic systems with extended replication and signaling," Microelectronics Advance Research Initiative (MEL-ARI), Technical Report 28667, 2001, 3rd Year Annual Report.
128. M. Stanisavljević, A. Schmid, and Y. Leblebici, "Optimization of the averaging reliability technique using low redundancy factors for nanoscale technologies," *IEEE Transactions on Nanotechnology*, vol. 8, no. 3, pp. 379–390, 2009.
129. M. K. Stojcev, G. L. Djordjevic, and M. D. Krstic, "A hardware mid-value select voter architecture," *Microelectronics Journal*, vol. 32, no. 1, pp. 149–162, 2001.
130. M. D. Krstic, M. K. Stojcev, G. L. Djordjevic, and I. D. Andrejic, "A mid-value select voter," *Microelectronics and Reliability*, vol. 45, no. 3-4, pp. 733–738, 2005.
131. G. Latif-Shabgahi, J. M. Bass, and S. Bennett, "Efficient implementation of inexact majority and median voters," *Electronics Letters*, vol. 36, no. 15, pp. 1326–1328, 2000.
132. G. Latif-Shabgahi and A. J. Hirst, "A fuzzy voting scheme for hardware and software fault tolerant systems," *Fuzzy Sets and Systems*, vol. 150, no. 3, pp. 579–598, 2005.
133. C. Chiou and T. C. Yang, "Self-purging redundancy with adjustable threshold for tolerating multiple module failures," *Electronics Letters*, vol. 31, no. 11, pp. 930–931, 1995.
134. G. Latif-Shabgahi, J. M. Bass, and S. Bennett, "History-based weighted average voter: a novel software voting algorithm for fault-tolerant computer systems," in *Proceedings of the 9th Euromicro Workshop on Parallel and Distributed Processing*, 7–9 Feb. 2001, pp. 402–409.
135. A. Schmid and Y. Leblebici, "Regular array of nanometer-scale devices performing logic operations with fault-tolerance capability," in *Proceedings of the 4th IEEE Conference on Nanotechnology (IEEE-NANO)*, 16–19 Aug. 2004, pp. 399–401.
136. A. Schmid and Y. Leblebici, "Realisation of multiple-valued functions using the capacitive threshold logic gate," *IEE Proceedings -Computers and Digital Techniques*, vol. 151, no. 6, pp. 435–447, 18 Nov. 2004.
137. A. Schmid and Y. Leblebici, "A highly fault tolerant PLA architecture for failure-prone nanometer CMOS and novel quantum device technologies," in *Proceedings of the 19th IEEE International Symposium on Defect and Fault Tolerance in VLSI Systems (DFT)*, 10–13 Oct. 2004, pp. 39–47.

138. E. Schuler and L. Carro, "Reliable digital circuits design using analog components," in *Proceedings of the 11th Annual IEEE International Mixed-Signals Testing Workshop (IMSTW)*, Cannes, France, June 2005, pp. 166–170.
139. A. Michels, L. Petroli, C. A. L. Lisboa, F. Kastensmidt, and L. Carro, "SET fault tolerant combinational circuits based on majority logic," in *Proceedings of the 21st IEEE International Symposium on Defect and Fault Tolerance in VLSI Systems (DFT)*, 4–6 Oct. 2006, pp. 345–352.
140. S. Aunet and V. Beiu, "Ultra low power fault tolerant neural inspired CMOS logic," in *Proceedings of the IEEE International Joint Conference on Neural Networks (IJCNN)*, vol. 5, 31 July–4 Aug. 2005, pp. 2843–2848.
141. S. Aunet, B. Oelmann, P. A. Norseng, and Y. Berg, "Real-time reconfigurable subthreshold CMOS perceptron," *IEEE Transactions on Neural Networks*, vol. 19, no. 4, pp. 645–657, April 2008.
142. S. Mitra, M. Zhang, N. Seifert, T. M. Mak, and K. S. Kim, "Soft error resilient system design through error correction," in *Proceedings of the IFIP International Conference on Very Large Scale Integration*, 16–18 Oct. 2006, pp. 332–337.
143. Y. M. Hsu and J. Swartzlander, E. E., "VLSI concurrent error correcting adders and multipliers," in *Proceedings of the IEEE International Workshop on Defect and Fault Tolerance in VLSI Systems (DFT)*, 27–29 Oct. 1993, pp. 287–294.
144. S. A. Al-Arian and M. B. Gumusel, "HPTR: Hardware partition in time redundancy technique for fault tolerance," in *Proceedings of the IEEE SoutheastCon*, 12–15 April 1992, pp. 630–633.
145. H. Al-Asaad and E. Czeck, "Concurrent error correction in iterative circuits by recomputing with partitioning and voting," in *Proceedings of the 11th Annual IEEE VLSI Test Symposium. Digest of Papers*, 6–8 April 1993, pp. 174–177.
146. W. L. Gallagher and J. Swartzlander, E. E., "Fault-tolerant Newton-Raphson and Goldschmidt dividers using time shared TMR," *IEEE Transactions on Computers*, vol. 49, no. 6, pp. 588–595, June 2000.
147. W. J. Townsend, J. A. Abraham, and J. Swartzlander, E. E., "Quadruple time redundancy adders [error correcting adder]," in *Proceedings of the 18th IEEE International Symposium on Defect and Fault Tolerance in VLSI Systems (DFT)*, 3–5 Nov. 2003, pp. 250–256.
148. A. J. KleinOsowski and D. J. Lilja, "The NanoBox project: Exploring fabrics of self-correcting logic blocks for high defect rate molecular device technologies," in *Proceedings of the IEEE Computer society Annual Symposium on VLSI*, 19–20 Feb. 2004, pp. 19–24.
149. M. Zhang and N. R. Shanbhag, "A CMOS design style for logic circuit hardening," in *Proceedings of the 43rd IEEE International Annual Reliability Physics Symposium*, 17–21 Apr. 2005, pp. 223–229.
150. Q. Zhou and K. Mohanram, "Cost-effective radiation hardening technique for combinational logic," in *Proceedings ot the IEEE/ACM International Conference on Computer Aided Design (ICCAD)*, 7–11 Nov. 2004, pp. 100–106.
151. Q. Zhou and K. Mohanram, "Gate sizing to radiation harden combinational logic," *IEEE Transactions on Computer-Aided Design of Integrated Circuits and Systems*, vol. 25, no. 1, pp. 155–166, Jan. 2006.
152. M. R. Choudhury, Q. Zhou, and K. Mohanram, "Design optimization for single-event upset robustness using simultaneous dual-VDD and sizing techniques," in *Proceedings of the IEEE/ACM International Conference on Computer-Aided Design (ICCAD)*, 5–9 Nov. 2006, pp. 204–209.
153. K. Mohanram and N. A. Touba, "Partial error masking to reduce soft error failure rate in logic circuits," in *Proceedings of the 18th IEEE International Symposium on Defect and Fault Tolerance in VLSI Systems (DFT)*, 3–5 Nov. 2003, pp. 433–440.
154. K. Mohanram, "Error detection and tolerance for scaled electronic technologies," in *Proceedings of the IEEE International Symposium on Defect and Fault Tolerance of VLSI Systems (DFT)*, 1–3 Oct. 2008, pp. 83–83.

155. M. R. Choudhury and K. Mohanram, "Approximate logic circuits for low overhead, non-intrusive concurrent error detection," in *Proceedings of the Design, Automation and Test in Europe (DATE)*, 10–14 Mar. 2008, pp. 903–908.
156. D. Clark, "Teramac: Pointing the way to real-world nanotechnology," *IEEE Computational Science & Engineering*, vol. 5, no. 3, pp. 88–90, 1998.
157. J. Lach, W. H. Mangione-Smith, and M. Potkonjak, "Low overhead fault-tolerant FPGA systems," *IEEE Transactions on Very Large Scale Integration (VLSI) Systems*, vol. 6, no. 2, pp. 212–221, June 1998.
158. D. Mange, M. Sipper, A. Stauffer, and G. Tempesti, "Toward robust integrated circuits: The embryonics approach," *Proceedings of the IEEE*, vol. 88, no. 4, pp. 516–543, April 2000.
159. M. Mishra and S. C. Goldstein, "Defect tolerance at the end of the roadmap," in *Proc. International Test Conference (ITC)*, vol. 1, Sept. 30–Oct. 2, 2003, pp. 1201–1210.
160. S. C. Goldstein, "The impact of the nanoscale on computing systems," in *Proceedings of the IEEE/ACM International Conference on Computer-Aided Design (ICCAD)*, 6–10 Nov. 2005, pp. 655–661.
161. G. Snider, P. Kueckes, and R. Williams, "CMOS-like logic in defective, nanoscale crossbars," *Nanotechnology*, vol. 15, pp. 881–891, 2004.
162. D. Bhaduri, S. Shukla, P. Graham, and M. Gokhale, "Comparing reliability-redundancy tradeoffs for two von Neumann multiplexing architectures," *IEEE Transactions on Nanotechnology*, vol. 6, no. 3, pp. 265–279, May 2007.
163. R. I. Bahar, D. Hammerstrom, J. Harlow, W. H. Joyner Jr., C. Lau, D. Marculescu, A. Orailoglu, and M. Pedram, "Architectures for silicon nanoelectronics and beyond," *Computers*, vol. 40, no. 1, pp. 25–33, 2007.
164. *Productive Nanosystems: A Technology Roadmap*, , K. E. Drexler, J. Randall, S. Corchnoy, A. Kawczak, M. L. Steve, (eds.). Palo Alto, CA: Battelle Memorial & Foresight Nanotech, 2007.
165. J. Allen M. Johnson and M. Malek, "Survey of software tools for evaluating reliability, availability, and serviceability," *ACM Computing Surveys*, vol. 20, no. 4, pp. 227–269, 1988.
166. R. L. Dobrushin and S. I. Ortyukov, "Lower bound for the redundancy of self-correcting arrangements of unreliable functional elements," *Problems of Information Transmission*, vol. 13, pp. 59–65, 1977.
167. R. L. Dobrushin and S. I. Ortyukov, "Upper bound for the redundancy of self-correcting arrangements of unreliable functional elements," *Problems of Information Transmission*, vol. 13, pp. 203–218, 1977.
168. R. Reischuk and B. Schmeltz, "Area efficient methods to increase the reliability of combinatorial circuits," in *STACS*, ser. Lecture Notes in Computer Science, vol. 349, B. Monien and R. Cori, (eds.). New York, NY: Springer, 1989, pp. 314–326.
169. R. Reischuk and B. Schmeltz, "Reliable computation with noisy circuits and decision trees-a general n log n lower bound," in *Proceedings of the 32nd Annual Symposium on Foundations of Computer Science*, 1–4 Oct. 1991, pp. 602–611.
170. R. Reischuk and B. Schmeltz, "Area efficient methods to increase the reliability of circuits," in *Data Structures and Efficient Algorithms*, ser. Lecture Notes in Computer Science, vol. 594. Springer, Berlin, Heidelberg, 1992, pp. 363–389.
171. T. Feder, "Reliable computation by networks in the presence of noise," *IEEE Transactions on Information Theory*, vol. 35, no. 3, pp. 569–571, May 1989.
172. B. Hajek and T. Weller, "On the maximum tolerable noise for reliable computation by formulas," *IEEE Transactions on Information Theory*, vol. 37, no. 2, pp. 388–391, Mar. 1991.
173. U. Feige, D. Peleg, P. Raghavan, and E. Upfal, "Computing with noisy information," *SIAM Journal on Computing*, vol. 23, pp. 1001–1018, 1994.
174. N. Pippenger, G. D. Stamoulis, and J. N. Tsitsiklis, "On a lower bound for the redundancy of reliable networks with noisy gates," *IEEE Transactions on Information Theory*, vol. 37, no. 3, pp. 639–643, May 1991.

175. W. Evans and N. Pippenger, "On the maximum tolerable noise for reliable computation by formulas," *IEEE Transactions on Information Theory*, vol. 44, no. 3, pp. 1299–1305, May 1998.
176. F. Martorell, A. Rubio, and S. Cotofana, "Analysis of the noise and parameter variations-tolerance of the averaging cell," in *Proceedings of the International Workshop on Design and Test of Defect-Tolerant Nanoscale Architectures (NANOARCH)*, 2005, pp. 1–6.
177. F. Martorell, S. D. Cotofana, and A. Rubio, "Fault tolerant structures for nanoscale gates," in *Proceedings of the 7th IEEE Conference on Nanotechnology (IEEE-NANO)*, 2–5 Aug. 2007, pp. 605–610.
178. F. Martorell and A. Rubio, "Cell architecture for nanoelectronic design," *Microelectronics Journal*, vol. 39, no. 8, pp. 1041–1050, 2008.
179. F. Martorell, S. D. Cotofana, and A. Rubio, "An analysis of internal parameter variations effects on nanoscaled gates," *IEEE Transactions on Nanotechnology*, vol. 7, no. 1, pp. 24–33, Jan. 2008.
180. D. Bhaduri and S. Shukla, "Nanolab-a tool for evaluating reliability of defect-tolerant nanoarchitectures," *IEEE Transactions on Nanotechnology*, vol. 4, no. 4, pp. 381–394, July 2005.
181. Hybrid automated reliability predictor (HARP). Duke and Clemson University. [Online]. Available: http://www.ee.duke.edu/~kst/software_packages.html
182. R. A. Sahner and K. S. Trivedi, "SHARPE: a modeler's toolkit," in *Proceedings of the IEEE International Computer Performance and Dependability Symposium*, 4–6 Sept. 1996, p. 58.
183. M. Boyd and S. Bavuso, "Simulation modeling for long duration spacecraft control systems," in *Proceedings of the Annual Reliability & Maintainability Symposium*, Atlanta, GA, USA, Jan. 1993, pp. 106–113.
184. Timenet. Technical University of Berlin. [Online]. Available: http://pdv.cs.tuberlin.de/~timenet/
185. Möbius. University of Illinois at Urbana-Champaign. [Online]. Available: www.mobius.uiuc.edu/index.html
186. SMART. University of California at Riverside. [Online]. Available: www.cs.ucr.edu/~ciardo/SMART/
187. J. B. Dugan, B. Venkataraman, and R. Gulati, "DIFtree: A software package for the analysis of dynamic fault tree models," in *Proceedings of the Annual Reliability and Maintainability Symposium*, 13–16 Jan. 1997, pp. 64–70.
188. D. Coppit and K. J. Sullivan, "Galileo: A tool built from mass-market applications," in *Proceedings of the International Conference on Software Engineering*, 4–11 June 2000, pp. 750–753.
189. Exelix. [Online]. Available: www.exelix.com/
190. S. Krishnaswamy, G. F. Viamontes, I. L. Markov, and J. P. Hayes, "Accurate reliability evaluation and enhancement via probabilistic transfer matrices," in *Proceedings of the Design, Automation and Test in Europe (DATE)*, 2005, pp. 282–287.
191. M. Kwiatkowska, G. Norman, D. Parker, and R. Segala, "Symbolic model checking of concurrent probabilistic systems using MTBDDs and simplex," School of Computer Science, University of Birmingham, Birmingham, UK, Technical Report CSR-99-01, 1999. [Online]. Available: http://www.cs.bham.ac.uk/~dxp/papers/CSR-99-01.pdf
192. D. Bhaduri and S. Shukla, "Tools and techniques for evaluating reliability of defect-tolerant nano architectures," in *Proceedings of the IEEE International Joint Conference on Neural Networks (IJCNN)*, vol. 4, 25–29 July 2004, pp. 2641–2646.
193. J. Srinivasan, "Lifetime reliability aware microprocessors," Ph.D. dissertation, University of Illinois at Urbana-Champaign, May 2006. [Online]. Available: http://rsim.cs.uiuc.edu/Pubs/srinivsn-phdthesis.pdf
194. G. Horton, "A new paradigm for the numerical simulation of stochastic Petri nets with general firing times," in *Proceedings of the European Simulation Symposium (ESS)*. Dresden, Germany, Oct. 2002.

195. S. Lazarova-Molnar, "The proxel-based method: Formalisation, analysis and applications," Ph.D. dissertation, Otto-von-Guericke University of Magdeburg, Germany, Nov. 2005. [Online]. Available: http://diglib.unimagdeburg.de/Dissertationen/2005/sanlazarova.pdf
196. M. Zhang and N. R. Shanbhag, "Soft-error-rate-analysis (sera) methodology," *IEEE Transactions on Computer-Aided Design*, vol. 25, no. 10, pp. 2140–2155, 2006.
197. B. Zhang, W.-S. Wang, and M. Orshansky, "Faser: Fast analysis of soft error susceptibility for cell-based designs," in *Proceedings of the 7th International Symposium Quality Electronic Design ISQED '06*, 2006.
198. R. R. Rao, K. Chopra, D. T. Blaauw, and D. M. Sylvester, "Computing the soft error rate of a combinational logic circuit using parameterized descriptors," *IEEE Transactions on Computer-Aided Design*, vol. 26, no. 3, pp. 468–479, Mar. 2007.
199. N. Miskov-Zivanov and D. Marculescu, "Mars-c: Modeling and reduction of soft errors in combinational circuits," in *Proceedings of the 43rd ACM/IEEE Design Automation Conference*, 2006, pp. 767–772.
200. N. Miskov-Zivanov and D. Marculescu, "Soft error rate analysis for sequential circuits," in *Proceedings of the Design, Automation & Test in Europe Conference & Exhibition DATE '07*, 2007, pp. 1–6.
201. P. Shivakumar, M. Kistler, S. W. Keckler, D. Burger, and L. Alvisi, "Modeling the effect of technology trends on the soft error rate of combinational logic," in *Proceedings of the International Conference Dependable Systems and Networks DSN 2002*, 2002, pp. 389–398.
202. S. Krishnaswamy, S. M. Plaza, I. L. Markov, and J. P. Hayes, "Signature-based ser analysis and design of logic circuits," *IEEE Transactions on Computer-Aided Design*, vol. 28, no. 1, pp. 74–86, 2009.
203. K. Patel, I. Markov, and J. Hayes, "Evaluating circuit reliability under probabilistic gate-level fault models," in *Proceedings of the International Workshop on Logic Synthesis (IWLS)*, Laguna Beach, CA, USA, May 2003, pp. 59–64.
204. V. Levin, "Probability analysis of combination systems and their reliability," *Engeering Cybernetics*, vol. 6, pp. 78–84, 1964.
205. W. Ibrahim, V. Beiu, and Y. A. Alkhawwar, "On the reliability of four full adder cells," in *Proceedings of the 4th International Conference on Innovations in Information Technology Innovations*, 18–20 Nov. 2007, pp. 720–724.
206. S. S. Ramani and S. Bhanja, "Any-time probabilistic switching model using Bayesian networks," in *Proceedings of the International Symposium on Low Power Electronics and Design (ISLPED)*, 2004, pp. 86–89.
207. T. Rejimon and S. Bhanja, "Scalable probabilistic computing models using Bayesian networks," in *Proceedings of the 48th Midwest Symposium on Circuits and Systems (MWSCAS)*, 7–10 Aug. 2005, pp. 712–715.
208. T. Rejimon and S. Bhanja, "An accurate probabilistic model for error detection," in *Proceedings of the 18th International Conference on VLSI Design*, 3–7 Jan. 2005, pp. 717–722.
209. T. Rejimon, K. Lingasubramanian, and S. Bhanja, "Probabilistic error modeling for nano-domain logic circuits," *IEEE Transactions on Very Large Scale Integration (VLSI) Systems*, vol. 17, no. 1, pp. 55–65, Jan. 2009.
210. J. Han, E. Taylor, J. Gao, and J. Fortes, "Faults, error bounds and reliability of nanoelectronic circuits," in *Proceedings of the 16th IEEE International Conference on Application-Specific Systems, Architecture Processors (ASAP)*, 23–25 July 2005, pp. 247–253.
211. E. Taylor, J. Han, and J. Fortes, "Towards accurate and efficient reliability modeling of nano-electronic circuits," in *Proceedings of the 6th IEEE Conference on Nanotechnology (IEEE-NANO)*, vol. 1, 17–20 June 2006, pp. 395–398.
212. M. R. Choudhury and K. Mohanram, "Accurate and scalable reliability analysis of logic circuits," in *Proceedings of the Design, Automation & Test in Europe Conference & Exhibition (DATE)*, 16–20 Apr. 2007, pp. 1–6.
213. D. T. Franco, M. C. Vasconcelosa, L. Navinera, and J.-F. Navinera, "Reliability analysis of logic circuits based on signal probability," in *Proceedings of the 15th IEEE International*

Conference on Electronics, Circuits and Systems (ICECS), Aug. 31 2008–Sept. 3 2008, pp. 670–673.
214. D. T. Franco, M. C. Vasconcelosa, L. Navinera, and J.-F. Navinera, "Signal probability for reliability evaluation of logic circuits," *Microelectronics Reliability*, vol. 48, no. 8-9, pp. 1586–1591, 2008.
215. S. Ercolani, M. Favalli, M. Damiani, P. Olivo, and B. Ricco, "Estimate of signal probability in combinational logic networks," in *Proceedings of the 1st European Test Conference*, 12–14 April 1989, pp. 132–138.
216. W. Ibrahim, V. Beiu, and M. H. Sulieman, "On the reliability of majority gates full adders," *IEEE Transactions on Nanotechnology*, vol. 7, no. 1, pp. 56–67, Jan. 2008.
217. Bsim3 homepage. [Online]. Available: http://www-device.eecs.berkeley.edu/~bsim3/intro.html
218. D. Bradley, C. Ortega-Sanchez, and A. Tyrrell, "Embryonics+immunotronics: A bio-inspired approach to fault tolerance," in *Proceedings of the 2nd NASA/DoD Workshop on Evolvable Hardware*, 13–15 July 2000, pp. 215–223.
219. S. Aunet, B. Oelmann, S. Abdalla, and Y. Berg, "Reconfigurable subthreshold CMOS perceptron," in *Proceedings of the IEEE International Joint Conference on Neural Networks (IJCNN)*, vol. 3, 25–29 July 2004, pp. 1983–1988.
220. S. Haykin, *Neural Networks: A Comprehensive Foundation*. Upper Saddle River, NJ: Prentice Hall, Inc., 1999.
221. K. Yamamori, S. Horiguchi, J. H. Kim, S. K. Park, and B. H. Ham, "The efficient design of fault-tolerant artificial neural networks," in *Proceedings of the IEEE International Conference on Neural Networks*, vol. 3, 27 Nov.–1 Dec. 1995, pp. 1487–1491.
222. P. Kerlirzin and P. Refregier, "Theoretical investigation of the robustness of multilayer perceptrons: Analysis of the linear case and extension to nonlinear networks," *IEEE Transactions on Neural Networks*, vol. 6, no. 3, pp. 560–571, May 1995.
223. R. C. Frye, E. A. Rietman, and C. C. Wong, "Back-propagation learning and nonidealities in analog neural network hardware," *IEEE Transactions on Neural Networks*, vol. 2, no. 1, pp. 110–117, Jan. 1991.
224. J. Nijhuis, B. Hofflinger, A. van Schaik, and L. Spaanenburg, "Limits to the fault-tolerance of a feedforward neural network with learning," in *Proceedings of the 20th International Symposium Fault-Tolerant Computing (FTCS). Digest of Papers*, 26–28 June 1990, pp. 228–235.
225. M. J. S. Smith, "An analog integrated neural network capable of learning the Feigenbaum logistic map," *IEEE Transactions on Circuits and Systems*, vol. 37, no. 6, pp. 841–844, June 1990.
226. A. Schmid and Y. Leblebici, "Robust circuit and system design methodologies for nanometer-scale devices and single-electron transistors," in *Proceedings of the 3rd IEEE Conference on Nanotechnology (IEEE-NANO)*, vol. 2, 12–14 Aug. 2003, pp. 516–519.
227. M. Stanisavljević, A. Schmid, and Y. Leblebici, "Analysis of reliability in nanoscale circuits and systems based on a-priori statistical fault-modeling methodology," in *Proceedings of the 48th Midwest Symposium on Circuits and Systems (MWSCAS)*, 7–10 Aug. 2005, pp. 1565–1568.
228. S. Aunet and M. Hartman, "Real-time reconfigurable threshold elements and some applications to neural hardware," in *Proceedings of the International Conference Evolvable System (ICES)*, Trondheim, Norway, Mar. 2003, pp. 365–376.
229. V. Beiu, S. Aunet, R. R. Rydberg III, A. Djupdal, and J. Nyathi, "The vanishing majority gate trading power and speed for reliability," in *Proceedings of the International Workshop on Design and Test of Defect-Tolerant Nanoscale Architectures (NANOARCH)*, 2005, pp. 1–8.
230. D. E. Rumelhart, G. E. Hinton, and R. J. Williams, *Parallel Distributed Processing: Explorations in the Microstructure of Cognition*. Volume 1: Foundations. Cambridge, MA: MIT Press, 1986, ch. Learning internal representations by error propagation, pp. 318–362.
231. G. Cauwenberghs, "Analog VLSI stochastic perturbative learning architectures," *Analog Integrated Circuits and Signal Processing*, vol. 13, no. 1–2, pp. 195–209, 1997.

232. A. Schmid and Y. Leblebici, "Robust circuit and system design methodologies for nanometer-scale devices and single-electron transistors," *IEEE Transactions on Very Large Scale Integration (VLSI) Systems*, vol. 12, no. 11, pp. 1156–1166, Nov. 2004.
233. M. Stanisavljević, A. Schmid, and Y. Leblebici, "Fault-tolerance of robust feed-forward architecture using single-ended and differential deep-submicron circuits under massive defect density," in *Proceedings of the International Joint Conference on Neural Networks (IJCNN)*, 16–21 July 2006, pp. 2771–2778.
234. L. Heller, W. Griffin, J. Davis, and N. Thoma, "Cascode voltage switch logic: A differential CMOS logic family," in *Proceedings of the IEEE International Solid-State Circuits Conference (ISSCC). Digest of Technical Papers*, vol. XXVII, Feb. 1984, pp. 16–17.
235. A. M. Ionescu, M. J. Declercq, S. Mahapatra, K. Banerjee, and J. Gautier, "Few electron devices: towards hybrid CMOS-SET integrated circuits," in *Proceedings of the 39th Design Automation Conference (DAC)*, 10–14 June 2002, pp. 88–93.
236. C. Lageweg, S. Cotofana, and S. Vassiliadis, "Achieving fanout capabilities in single electron encoded logic networks," in *Proceedings of the 6th International Conference on Solid-State and Integrated-Circuit Technology*, vol. 2, 22–25 Oct. 2001, pp. 1383–1386.
237. J. R. Tucker, "Complementary digital logic based on the "Coulomb blockade," *Journal of Applied Physics*, vol. 72, no. 9, pp. 4399–4413, 1992.
238. M. M. Ziegler and M. R. Stan, "CMOS/nano co-design for crossbar-based molecular electronic systems," *IEEE Transactions on Nanotechnology*, vol. 2, no. 4, pp. 217–230, Dec. 2003.
239. S. Mahapatra, V. Pott, S. Ecoffey, A. Schmid, C. Wasshuber, J. W. Tringe, Y. Leblebici, M. Declercq, K. Banerjee, and A. M. Ionescu, "SETMOS: A novel true hybrid SET-CMOS high current Coulomb blockade oscillation cell for future nano-scale analog ICs," in *Proceedings of the IEEE International Electron Devices Meeting (IEDM). Technical Digest*, 8–10 Dec. 2003, pp. 29.7.1–29.7.4.
240. H. Iwamura, M. Akazawa, and Y. Amemiya, "Single-electron majority logic circuits," *IEICE Transactions on Electronics*, vol. E81-C, pp. 42–48, 1998.
241. M. Sulieman and V. Beiu, "Design and analysis of SET circuits: using MATLAB modules and SIMON," in *Proceedings of the 4th IEEE Conference on Nanotechnology (IEEE-NANO)*, 16–19 Aug. 2004, pp. 618–621.
242. C. Wasshuber, H. Kosina, and S. Selberherr, "SIMON-a simulator for single-electron tunnel devices and circuits," *IEEE Transactions on Computer-Aided Design of Integrated Circuits and Systems*, vol. 16, no. 9, pp. 937–944, Sept. 1997.
243. R. I. Bahar, J. Mundy, and J. Chen, "A probabilistic-based design methodology for nanoscale computation," in *Proceedings of the International Conference on Computer Aided Design (ICCAD)*, 9–13 Nov. 2003, pp. 480–486.
244. R. I. Bahar, J. Chen, and J. Mundy, *Nano, Quantum and Molecular Computing: Implications to High Level Design and Validation*. Norwell, MA: Kluwer Academic Publishers, 2004, ch. A probabilistic-based design for nanoscale computation, pp. 133–156.
245. F. Martorell and A. Rubio, "Defect and fault tolerant cell architecture for feasible nanoelectronic designs," in *Proceedings of the International Conference on Design and Test of Integrated Systems in Nanoscale Technology (DTIS)*, 5–7 Sept. 2006, pp. 244–249.
246. A. Papoulis and S. U. Pillai, *Probability, Random Variables and Stochastic Processes*, 4th ed., New York, NY: McGraw-Hill, 2002.
247. S. Mitra, N. R. Saxena, and E. J. McCluskey, "Common-mode failures in redundant VLSI systems: A survey," *IEEE Transactions on Reliability*, vol. 49, no. 3, pp. 285–295, Sept. 2000.
248. M. R. Choudhury and K. Mohanram, "Reliability analysis of logic circuits," *IEEE Transactions on Computer-Aided Design of Integrated Circuits and Systems*, vol. 28, no. 3, pp. 392–405, Mar. 2009.
249. M. C. Hansen, H. Yalcin, and J. P. Hayes, "Unveiling the ISCAS-85 benchmarks: A case study in reverse engineering," *IEEE Design & Test of Computers*, vol. 16, no. 3, pp. 72–80, July–Sept. 1999.

250. G. Karypis, R. Aggarwal, V. Kumar, and S. Shekhar, "Multilevel hypergraph partitioning: Applications in VLSI domain," *IEEE Transactions on Very Large Scale Integration (VLSI) Systems*, vol. 7, no. 1, pp. 69–79, 1999.
251. S. Yang, "Logic synthesis and optimization benchmarks user guide," Microelectronic Center of North Carolina, Technical Report 1/95, 1991.
252. G. Karypis and V. Kumar, "hMeTiS: A hypergraph partitioning package," Department of Computer Science and Engineering, University of Minnesota, Minneapolis, Technical Report, 1998.
253. R. Rajaraman and D. F. Wong, "Optimal clustering for delay minimization," in *Proceedings of the 30th Conference on Design Automation (DAC)*, 14–18 June 1993, pp. 309–314.
254. H. H. Yang and D. F. Wong, "Circuit clustering for delay minimization under area and pin constraints," *IEEE Transactions on Computer-Aided Design of Integrated Circuits and Systems*, vol. 16, no. 9, pp. 976–986, Sept. 1997.
255. R. Murgai, R. K. Brayton, and A. Sangiovanni-Vincentelli, "On clustering for minimum delay/area," in *Proceedings of the IEEE International Conference on Computer-Aided Design (ICCAD). Digest of Technical Papers*, 11–14 Nov. 1991, pp. 6–9.
256. E. L. Lawler, K. N. Levitt, and J. Turner, "Module clustering to minimize delay in digital networks," *IEEE Transactions on Computers*, vol. C-18, no. 1, pp. 47–57, Jan. 1969.
257. C. Ababei and K. Bazargan, "Timing minimization by statistical timing hMetis-based partitioning," in *Proceedings of the 16th International Conference on VLSI Design*, 4–8 Jan. 2003, pp. 58–63.
258. C. M. Fiduccia and R. M. Mattheyses, "A linear-time heuristic for improving network partitions," in *Proceedings of the 19th Design Automation Conference (DAC)*, 14–16 June 1982, pp. 175–181.
259. S.-L. Ou and M. Pedram, *Timing-driven Partitioning Using Iterative Quadratic Programming*, 2001, see "Coming Attractions!". [Online]. Available: http://atrak.usc.edu/~massoud/
260. M. Shih and E. S. Kuh, "Quadratic boolean programming for performance-driven system partitioning," in *Proceedings of the 30th Design Automation Conference (DAC)*, 14–18 June 1993, pp. 761–765.
261. J. Cong and C. Wu, "Global clustering-based performance-driven circuit partitioning," in *Proceedings of the International Symposium on Physical Design (ISPD)*, 2002, pp. 149–154.
262. J. Minami, T. Koide, and S. Wakabayashi, "A circuit partitioning algorithm under path delay constraints," in *Proceedings of the IEEE Asia-Pacific Conference on Circuits and Systems (IEEE APCCAS)*, 24–27 Nov. 1998, pp. 113–116.
263. S. Wakabayashi, "An iterative improvement circuit partitioning algorithm under path delay constraints," *IEICE Transactions on Fundamentals of Electronics, Communications and Computer Sciences. Special Section on VLSI Design and CAD Algorithms*, vol. 83, no. 12, pp. 2569–2576, 2000.
264. Z. Chishti and T. N. Vijaykumar, "Optimal power/performance pipeline depth for SMT in scaled technologies," *IEEE Transactions on Computers*, vol. 57, no. 1, pp. 69–81, Jan. 2008.
265. S. Aunet, Y. Berg, and V. Beiu, "Ultra low power redundant logic based on majority-3 gates," in *Proceedings of the IFIP VLSI-SoC*, Perth, Australia, Oct. 2005, pp. 553–558.

Index

GPSR Compliance

The European Union's (EU) General Product Safety Regulation (GPSR) is a set of rules that requires consumer products to be safe and our obligations to ensure this.

If you have any concerns about our products, you can contact us on ProductSafety@springernature.com

In case Publisher is established outside the EU, the EU authorized representative is:

Springer Nature Customer Service Center GmbH
Europaplatz 3
69115 Heidelberg, Germany

Batch number: 10584688

Printed by Printforce, the Netherlands